Lecture Notes in Mathematics

A collection of informal reports and seminars
Edited by A. Dold, Heidelberg and B. Eckmann, Zürich

293

Ronald A. DeVore
Oakland University, Rochester, MI/USA

The Approximation of Continuous Functions by Positive Linear Operators

Springer-Verlag
Berlin · Heidelberg · New York 1972

AMS Subject Classifications (1970): 41-02, 41 A 10, 41 A 25, 41 A 35, 42 A 04, 42 A 08

ISBN 3-540-06038-3 Springer-Verlag Berlin · Heidelberg · New York
ISBN 0-387-06038-3 Springer-Verlag New York · Heidelberg · Berlin

Offsetdruck: Julius Beltz, Hemsbach/Bergstr.

PREFACE

These notes study linear methods of approximation which are given by a sequence (L_n) of positive linear operators. The essential ingredient being that of positivity. The main theme is to relate the smoothness of the function f being approximated with the rate of decrease of $||f - L_n(f)||$. This is accomplished in the usual setting of direct theorems, inverse theorems, saturation and approximation of classes of functions.

The fundamental ideas involved in direct estimates can be found in the pioneering book of P.P. Korovkin[4] and several of the more recent textbooks on approximation. However, most of the material appears in "book form" here for the first time. The main exception being the results on approximation by positive convolution operators, which have considerable overlap with the recent book of P.L. Butzer and R.J. Nessel[1].

I have written the notes at a level which presupposes a knowledge of the fundamental aspects of approximation theory, especially as pertains to the degree of approximation. Most of the necessary background material can be found in the now classic book of G.G. Lorentz[2]. For a good understanding of the material developed here for convolution operators, I expect that the reader will have to make several excursions into Butzer and Nessel[1].

The notes concentrate solely on spaces of continuous function (periodic and non-periodic) on a finite interval. I have not developed the theory for L_p - spaces since I know of little in these spaces that goes beyond what is already contained in Butzer and Nessel[1]. The only examples considered are those which fall comfortably in the grasps of the general theory. However, I believe that the reader will find that most of the better known examples are covered.

ACKNOWLEDGEMENT. I would like to thank Professor P.L. Butzer and his group at Aachen, most notably Professor R.J. Nessel and Dr.'s E. Görlich, E. Stark, K. Scherer, who were generous enough to comment on a rough first draft of these notes which I sent them in the spring of 1971. During the academic year 1971-1972, I had the opportunity to lecture on this material in a seminar on approximation theory at the University of Alberta. This effected considerably the final version of the notes, and I would like to thank all those who participated in this seminar, especially Professors A. Meir and A. Sharma who invited me to spend the year in Alberta. Finally, I would like to thank Mrs. Dorothy Farren for typing the first draft and Miss Olwyn Buckland for typing the final version of the manuscript.

Edmonton, June, 1972

CONTENTS

CHAPTER 1

PRELIMINARIES

1.1. Introduction. This first chapter will be devoted to stating, generally without proof, some fundamental results from approximation theory and Fourier analysis. Proofs can usually be found in the classical books on the appropriate subject. The material on approximation of functions can be found in the book of G.G. Lorentz$_2$, while the treatise of A. Zygmund$_1$ should be referred to for results from Fourier analysis. There is some middle ground material between approximation and Fourier analysis, which is best found in the work of P.L. Butzer and R.J. Nessel$_1$. We also quote some results on Chebyshev systems from S. Karlin and W.J. Studden$_1$, and orthogonal polynomials from G. Szegö$_1$. In the rare instance, that there is no reference for the exact result we need, a proof will be supplied.

In some cases, proofs of results stated in this chapter can actually be found in subsequent chapters. For example, Jackson's Theorem is proved in Chapter 2 and the Bernstein Inverse Theorem follows easily from the material in Chapter 8. However, we do not strive for this kind of completeness.

Most of our notation and terminology will be introduced in this chapter. Later in the text, when there may be some question as to notation, we will generally give reference back to the original usage.

1.2. Chebyshev Systems. $C[a,b]$ denotes the space of continuous functions on $[a,b]$ and $C^* = C^*[-\pi,\pi]$, the space of 2π-periodic and continuous functions on the line. Both of these spaces are equipped with the supremum norm over the appropriate interval. For example, for $f \in C[a,b]$

$$||f|| = \sup_{a \le x \le b} |f(x)|$$

We will use the notation $\|\cdot\|[a,b]$ to indicate that the supremum is taken over $[a,b]$ whenever it is necessary to make it clear which interval the norm is taken over.

A set of functions $\{u_0,u_1, \ldots,u_n\} \subseteq C[a,b]$ is called a Chebyshev system on $[a,b]$ if each function $u \in sp(u_0, \ldots,u_n)$ has at most n zeros unless it is identically zero on $[a,b]$. For the periodic case, $\{u_0, \ldots,u_n\} \subseteq C^*$ and any $u \in sp(u_0, \ldots,u_n)$ has at most n zeros on $[-\pi,\pi)$, unless it is identically zero. The space $U_n = sp(u_0, \ldots,u_n)$ is called a <u>Chebyshev space</u>. P_n, the space of algebraic polynomials of degree $\leq n$ is a Chebyshev space for any $[a,b]$ and T_n the space of trigonometric polynomials of degree $\leq n$ is a Chebyshev subspace of C^*.

A Chebyshev space on $[a,b]$ can also be characterized by the following interpolation property.

1.2.1. <u>If</u> $a \leq x_0 < x_1 < \ldots < x_n \leq b$ <u>and</u> $(y_i)_{i=0}^n$ <u>are real numbers then there is a unique</u> $u \in sp(u_0, \ldots,u_n)$ <u>such that</u>

$$u(x_i) = y_i \qquad i = 0,1, \ldots n .$$

If U_n is a Chebyshev subspace of $C[a,b]$ with each non-zero $u \in U_n$ having n continuous derivatives on $[a,b]$ and at most n zeros on $[a,b]$ <u>counting multiplicity</u> then we say U_n is an <u>extended Chebyshev space</u>. If for each $k = 0,\ldots,n$, $sp(u_0,\ldots,u_k)$ is an extended Chebyshev system, then we say U_n is an <u>extended complete Chebyshev space</u>. For an extended complete Chebyshev space there is a canoncial basis $\{u_0,\ldots,u_n\}$ described by

1.2.2. $$u_0(t) = w_0(t)$$

$$u_1(t) = w_0(t)\int_a^t w_1(x_1)dx_1$$

$$\cdot$$

$$u_n(t) = w_0(t) \int_a^t w_1(x_1) \int_a^{x_1} w_2(x_2) \cdots \int_a^{x_{n-1}} w_n(x_n)dx_n \cdots dx_1$$

where each w_i is a continuous strictly positive function on $[a,b]$, $w_i \varepsilon C^{(n-i)}[a,b]$.

Let $f \in C[a,b]$ (C^*) and U_n be a Chebyshev subspace of $C[a,b]$ (C^*). A function $u^* \in U_n$ is called a best approximation to f from U_n if

$$\| f-u^* \| = \inf_{u \in U_n} \|f-u\|$$

The following theorem of Chebyshev gives the existence, uniqueness and characterization of u^*.

Theorem 1.1. Let $f \varepsilon C[a,b]$ $(C^*[-\pi,\pi])$ and $\{u_0, \ldots, u_n\}$ be a Chebyshev system on $[a,b]([-\pi,\pi])$. Then the best approximation u^* to f exists and is unique. If u is any function in $sp(u_0, \ldots, u_n)$ then u is the best approximation to f if and only if there exist points

$a \leq x_0 < x_1 < \ldots < x_{n+1} \leq b(-\pi \leq x_0 < x_1 < \ldots < x_{n+1} < \pi)$ such that

$f-u$ alternately takes on the values $\pm \| f-u \|$ at the points x_i, $i = 0, \ldots, n+1$.

1.3. Classes of Functions. There are various methods for measuring the smoothness of functions. The divided difference operator $\Delta_t(f,x)$ is defined for $t \geq 0$ by

$$\Delta_t(f,x) = f(x+t) - f(x)$$

If $f \in C[a,b]$, then the modulus of continuity of f is defined by

$$\omega(f,h) = \sup_{0 \le t \le h} \| \Delta_t(f,x) \| [a,b-t] \quad 0 \le h \le b-a$$

where the norm is taken with respect to the variable x. A similar definition holds for $f \in C^*$ except now the norm can be taken over the whole line. The modulus of continuity has the following fundamental properties:

1.3.1. $\omega(f,t)$ <u>is non-decreasing</u>, <u>continuous</u>, <u>and</u> $\omega(f,0) = 0$.

1.3.2. <u>If</u> $t_1,t_2 > 0$, <u>then</u> $\omega(f,t_1+t_2) \le \omega(f,t_1) + \omega(f,t_2)$.

1.3.3. <u>If</u> $\lambda > 0$ <u>and</u> $t > 0$, <u>then</u> $\omega(f,\lambda t) \le (\lambda+1)\omega(f,t)$.

1.3.4. <u>If</u> $t_1 < t_2$ <u>then</u> $t_2^{-1}\omega(f,t_2) \le 2t_1^{-1}\omega(f,t_1)$.

The conditions (1.3.1.) and (1.3.2.) characterize the modulus of continuity in the sense that if ω is any function which has these two properties, then ω is a modulus of continuity of a function in $C[a,b]$ (namely $\omega(t-a)$).

When ω is a modulus of continuity, we denote by $C_\omega(M)$ the set of all functions f for which

$$\omega(f,t) \le M\omega(t) \qquad 0 \le t \le b-a$$

The notation C_ω stands for the union of all the $C_\omega(M)$, $M > 0$. Sometimes, we will need to have a compact set and so we will restrict the norms of the functions by $\| f \| \le M_0$ and denote this new class by $C_\omega(M,M_0)$.

The most important case here is when $\omega(t) = t^\alpha$, with $0 < \alpha \le 1$, which are called the Lipschitz α classes. The common notation is $\text{Lip}\,\alpha = C_{t^\alpha}$. Similar notations, $\text{Lip}(\alpha,M)$ and $\text{Lip}(\alpha,M,M_o)$ are used for $C_{t^\alpha}(M)$ and $C_{t^\alpha}(M,M_o)$.

For higher smoothness, we let $W^{(r)}(\alpha,M,M_o,\ldots,M_r)$ denote the set of functions for which

$$\|f^{(i)}\| \le M_i \qquad i = 0,1,\ldots,r$$

$$\omega(f^{(r)},t) \le Mt^\alpha \qquad 0 \le t \le b-a$$

When we do not restrict the norms of the $f^{(i)}$ then we denote this class by $W^{(r)}(\alpha,M)$. The class $W^{(r)}(\alpha)$ is the union of all the classes $W^{(r)}(\alpha,M), M > 0$.

We denote by $L_\infty^{(r)}(M)$, the class of all functions f whose r^{th} derivative is $\le M$ a.e.. $L_\infty^{(r)}$ is the union of all the $L_\infty^{(r)}(M)$. The classes $L_\infty^{(r)}$ and $W^{(r-1)}(1)$ are the same.

We can also get at higher smoothness by using higher order divided differences. The rth order divided difference Δ_t^r is defined as the r-fold composition of Δ_t with itself. The rth order modulus of continuity is then given by

$$\omega_r(f,h) = \sup_{0\le t\le h} \|\Delta_t^r(f,x)\| \ [a,b-rt] \qquad 0 \le h \le \frac{b-a}{r}$$

Particularly important is the second order modulus of continuity since its behaviour cannot be characterized in terms of the first order modulus of continuity.

If $0 < \alpha \le 2$, we define the class $\mathrm{Lip}^*(\alpha,M)$ as the collection of all functions $f \in C[a,b]$ $(C^*[-\pi,\pi])$ such that

$$\omega_2(f,t) \le 2M\, t^\alpha .$$

$Lip^*\alpha$ is the union of all $Lip^*(\alpha,M)$ over all M's . For $0 < \alpha < 1$, the classes $Lip^*\alpha$ and Lip α are the same. In fact, there are constants M_1,M_2 and M_3 depending on α , such that

$$Lip(\alpha,M_1) \subseteq Lip^*(\alpha,M_2) \subseteq Lip(\alpha,M_3) \qquad (1.3.5)$$

Similarly, for $1 < \alpha < 2$, the classes $Lip^*\alpha$ and $W^{(1)}(\alpha)$ are the same. Again a relation like (1.3.5.) holds. When $\alpha = 2$, we can say more

$$Lip^*(2,M) = W^{(1)}(1,2M) \qquad (1.3.6).$$

When $\alpha = 1$, the class Lip 1 is properly contained in Lip^*1 . The class Lip^*1 is commonly referred to as the class of quasi-smooth functions, also called the Zygmund class. The class of smooth functions S consists of all functions f such that

$$\omega_2(f,t) = o(t) .$$

Given a function in C[a,b] , it is sometimes necessary to define f outside [a,b] in such a way that as much smoothness as possible can be retained. It is always possible to extend f to get a new function g defined on the entire line (see $Timan_1$, p. 121) with

$$\omega_2(g,t) \leq 5\omega_2(f,t) \qquad 0 \leq t \leq \tfrac{1}{2}(b-a) \qquad (1.3.7.)$$

where now in the definition of $\omega_2(g,t)$ the norm is taken over all $-\infty < x < \infty$.

1.4. Fourier Series. If $f \in C^*$, the complex Fourier coefficients of f are defined by

$$\hat{f}(k) = \frac{1}{2\pi}\int_{-\pi}^{\pi} f(t)e^{-ikt}dt \qquad k = 0, \pm 1, \ldots$$

and the real Fourier coefficients by

$$a_k(f) = \frac{1}{\pi}\int_{-\pi}^{\pi} f(t)\cos kt\,dt\ , \quad b_k(f) = \frac{1}{\pi}\int_{-\pi}^{\pi} f(t)\sin kt, \quad k = 0,1,2, \ldots .$$

Thus, the Fourier series of f is

$$f \sim \sum_{k-\infty}^{\infty} \hat{f}(k)e^{ikx} = \frac{a_0(f)}{2} + \sum_{1}^{\infty} a_k(f)\cos kx + b_k(f)\sin kx = \sum_{0}^{\infty} A_k(f,x)$$

Related to this is the conjugate series

$$\sum_{1}^{\infty}(-b_k(f)\cos kx + a_k(f)\sin kx) \qquad (1.4.1.)$$

If $f \in C^*$ then (1.4.1.) is the Fourier series of a function $g \in L_1[-\pi,\pi]$. We call g the conjugate function of f and write $g = \tilde{f}$.

The nth partial sum of the Fourier series of f is denoted by

$$S_n(f,x) = \sum_{0}^{n} A_k(f,x) .$$

Similarly, for a Borel measure $d\mu$ defined on $[-\pi,\pi)$, we define the complex Fourier coefficients of $d\mu$ by

$$\check{\mu}(k) = \frac{1}{2\pi}\int_{-\pi}^{\pi} e^{-ikt}d\mu(t)\ , \qquad k = 0, \pm 1, \ldots$$

In the case that $d\mu$ is an even measure, then the sine terms drop and

$$d\mu \sim \sum_{-\infty}^{\infty} \check{\mu}(k)e^{ikx} = \frac{\rho_0}{2} + \sum_{1}^{\infty} \rho_k \cos kx$$

where

$$\rho_k = \frac{1}{\pi} \int_{-\pi}^{\pi} \cos kt \, d\mu(t)$$

We call the ρ_k's the real Fourier coefficients of $d\mu$.

If f and g are in $C^*[-\pi,\pi]$ then the convolution of f with g is

$$(f*g)(x) = \frac{1}{\pi} \int_{-\pi}^{\pi} f(t)\, g(x-t)dt \tag{1.4.2.}$$

The function $f*g$ is also in $C^*[-\pi,\pi]$ and has the Fourier series

$$f*g \sim 2 \sum_{-\infty}^{\infty} \hat{f}(k)\, \hat{g}(k)e^{ikx} \tag{1.4.3.}$$

More generally we can define the convolution of f with a measure $d\mu$ by replacing $g(x-t)dt$ by $d\mu(x-t)$ in (1.4.2.) . Then $f*d\mu$ is in $C^*[-\pi,\pi]$ and (1.4.3.) also holds with $d\mu$ in place of g . In particular if $d\mu$ is even then

$$f*d\mu \sim \sum_{0}^{\infty} \rho_k A_k(f,x) \tag{1.4.4.}$$

It is important to point out that the factor 2 that appears in (1.4.3.) arises since we used $\frac{1}{\pi}$ rather than $\frac{1}{2\pi}$ in (1.4.2.) . The only reason for doing this is that the formula (1.4.4.) comes out in a more convenient form.

If A and B are two classes of 2π- periodic functions and L is a continuous, translation invariant operator mapping A into B, we say that L is a multiplier from A into B. We denote the set of all multipliers from A to B by (A,B). The classes (C^*,C^*) and (L_∞, L_∞) are the same and each L in (C^*,C^*) can be represented as convolution with a Borel measure $d\mu$ on $[-\pi,\pi)$;

$$L(f) = f*d\mu$$

The effect of L is to take the function f and map it into the function whose Fourier series is

$$2\sum_{-\infty}^{\infty} \check{\mu}(k)\ \hat{f}(k)e^{ikx} \qquad (1.4.5.)$$

We usually say that $(\check{\mu}(k))_{-\infty}^{\infty}$ is the multiplier. In the case that $d\mu$ is even then (1.4.5.) becomes

$$\sum_{0}^{\infty} \rho_k A_k(f,x)$$

and we say $(\rho_k)_0^\infty$ is a multiplier in (C^*,C^*). Thus the condition that $(\rho_k)_0^\infty$ is in (C^*,C^*) is that it be the real Fourier-Stieljes coefficients of an even Borel measure on $[-\pi,\pi)$. The norm of a multiplier is it's operator norm. In the above case

$$||L|| = \frac{1}{\pi}\int_{-\pi}^{\pi} |d\mu|$$

Given a sequence $(\lambda_k)_{k=0}^{\infty}$, it is usually difficult to determine whether it is a real Fourier-Stieljes sequence. A necessary and sufficient criteria is that the (C,1) means of the series $\frac{1}{2}\lambda_0 + \sum_1^{\infty} \lambda_k \cos kx$ be bounded in L_1 . That is, $(\lambda_k) \in (C^*,C^*)$ if and only if

$$\frac{1}{\pi}\int_{-\pi}^{\pi} |\tfrac{1}{2}\lambda_0 + \sum_1^n (1 - \frac{k}{n+1})\lambda_k \cos kx|\,dx = \frac{1}{\pi}\int_{-\pi}^{\pi} |(\tfrac{1}{2}\lambda_0 + \sum_1^n \lambda_k \cos kx)*F_n|\,dx = O(1) \tag{1.4.6.}$$

where F_n is the Fejér kernel

$$F_n(t) = \frac{1}{2(n+1)}\left(\frac{\sin(\frac{n+1}{2}t)}{\sin\frac{t}{2}}\right)^2$$

A sufficient condition that $(\lambda_k)_0^{\infty} \in (C^*,C^*)$ is that it is convex (i.e. $\Delta^2\lambda_k = \lambda_k - 2\lambda_{k+1} + \lambda_{k+2} \geq 0$) or more generally quasi-convex (i.e. $\sum_0^{\infty} (k+1)\,|\Delta^2\lambda_k| < +\infty$) . For in this case, applying summation by parts twice

$$\tfrac{1}{2}\lambda_0 + \sum_1^n \lambda_k \cos kx = \sum_0^{n-2} (k+1)\Delta^2\lambda_k F_k(x) + n\Delta\lambda_{n-1}F_{n-1}(x) + \lambda_n D_n(x) \tag{1.4.7.}$$

with D_n the Dirichlet kernel

$$D_n(t) = \frac{1}{2} + \sum_1^n \cos kt = \frac{\sin(2n+1)(\frac{t}{2})}{2\sin\frac{t}{2}}$$

When $(\lambda_k)_0^{\infty}$ is quasi-convex, it is known that $n\Delta\lambda_{n-1} \to 0$ and (λ_n) is bounded, so that convoluting (1.4.7.) with F_n , we have

$$\frac{1}{\pi}\int_{-\pi}^{\pi} |\tfrac{1}{2}\lambda_0 + \sum_1^n (1- \frac{k}{n+1})\lambda_k \cos kt \mid dt \leq \sum_0^{\infty} (k+1)\,|\Delta^2\lambda_k| + |n\Delta\lambda_{n-1}| + |\lambda_n| \tag{1.4.8.}$$

which shows that (1.4.6.) holds. The right hand side of (1.4.8.) can also be used to show that $\sum_{0}^{\infty} (k+1)\ |\Delta^2\lambda_k| + \sup|\lambda_n|$ is a bound for the norm of the multiplier $(\lambda_k)_0^{\infty}$.

<u>1.5. Degree of Approximation of Functions.</u> When $f \in C^*$ we define $E_n^*(f)$, the error in approximating f by elements of $\mathcal{T}_n$ by

$$E_n^*(f) = \inf\{\ \|f-T\| : T \in \mathcal{T}_n\}$$

We define $E_n(f)$, the error in approximation $f \in C[a,b]$ by elements of $\mathcal{P}_n$ in the same way.

One of the central themes in the theory of approximation of functions is to connect the smoothness of the function f with the rate of convergence of $E_n^*(f)$ to 0 . The pioneering result is the following theorem of D. Jackson.

<u>Theorem 1.2. For the periodic case, there is a constant</u> A <u>such that if</u> $f \in C_\omega(M)$ <u>then for</u> $n \geq 1$

$$E_n^*(f) \leq AM\omega(n^{-1}) \tag{1.5.1.}$$

<u>For each</u> $r \in N$ [i)], <u>there is a constant</u> A_r <u>such that whenever</u> $f \in W^{(r)}(\alpha,M)$ <u>then for</u> $n \geq r$

$$E_n^*(f) \leq A_r M\ n^{-r-\alpha} \tag{1.5.2.}$$

<u>The constants are independent of</u> f <u>and</u> n .

In the opposite direction, we have the following theorem of S. Bernstein.

i) N denotes the set of natural numbers, $N = \{1,2,3,...\}$.

<u>Theorem 1.3.</u> <u>If</u> $0 < \alpha < 1$, $r \in N$, <u>and</u> $f \in C^*$ <u>with</u>

$$E_n^*(f) = O(n^{-r-\alpha})$$

<u>then</u> $f \in W^{(r)}(\alpha)$.

Conspicuously absent is the case $\alpha = 1$ because in this case it is necessary to consider second order differences. Here, the proper characterization was given by A. Zygmund.

<u>Theorem 1.4.</u> <u>If</u> $r \in N$, <u>and</u> $f \in C^*$ <u>then</u>

$$E_n^*(f) = O(n^{-r})$$

<u>if and only if</u> $f^{(r-1)} \in Lip^* 1$. <u>Also,</u>

$$E_n^*(f) = o(n^{-r})$$

<u>if and only if</u> $f^{(r-1)}$ <u>is smooth on</u> $[-\pi,\pi]$.

Actually Theorems 1.3 and 1.4 can be derived from the following more complete results of S.B. Stečkin which gives an estimate for smoothness in terms of $E_n^*(f)$.

<u>Theorem 1.5</u> <u>If</u> $f \in C^*$ <u>and</u> $r \in N$ <u>then</u>

$$\omega_r(f,h) \leq A_r h^r \sum_{0}^{[h^{-1}]} (n+1)^{r-1} E_n^*(f) \quad 0 < h \leq 1 \qquad (1.5.3.)$$

The algebraic case is considerably different. The direct theorems of Jackson type hold equally well but the inverse theorems of Bernstein type do not hold in the above form. The missing ingredient is that approximation by algebraic polynomials is more efficient near the end points of the interval. This was first shown by A.F. Timan.

The following direct theorem is a contribution of several authors: A.F. Timan, V.K. Dzjadyk, G. Freud and J.A. Brudnyi .

Theorem 1.6. Let f be r times continuously differentiable on $[-1,1]$. If $k \in N$ then there is a constant $A_{k,r}$ such that for any $n \geq r+k+1$ we can find $P_n \in P_n$ with

$$|f(x) - P_n(x)| \leq A_{k,r} (\Delta_n(x))^r \omega_k(f^{(r)},\Delta_n(x)) \qquad -1 \leq x \leq 1 \tag{1.5.4.}$$

where $\Delta_n(x) = n^{-1}(1-x^2)^{\frac{1}{2}} + n^{-2}$, $n \in N$.

The inverse theorem to Theorem 1.6. was given by V.K. Dzjadyk and A.F. Timan.

Theorem 1.7. Let k , $r \in N$ and ω be a kth order modulus of continuity. Suppose f is a function in $C[-1,1]$ for which there exists a sequence of polynomials (P_n) , $P_n \in P_n$, with

$$|f(x) - P_n(x)| \leq (\Delta_n(x))^r \omega(\Delta_n(x)) \qquad -1 \leq x \leq 1 \tag{1.5.5.}$$

Then for $r = 0$,

$$\omega_k(f,h) \leq A_k h^k \int_h^1 u^{-k-1} \omega(u)du \qquad 0 \leq h \leq \frac{1}{2} \tag{1.5.6.}$$

For $r \geq 1$, if $\int_0^1 u^{-1}\omega(u)du < +\infty$ then

$$\omega_k(f^{(r)},h) \leq A_{k,r}[h^k \int_h^1 u^{-k-1}\omega(u)du + \int_0^h u^{-1}\omega(u)du] \quad 0 \leq h \leq \tfrac{1}{2} \qquad (1.5.7.)$$

It is important to note that if $E_n(f) = O(n^{-\alpha})$, on $[-1,1]$ $0 < \alpha \leq 2$, we can conclude that $f \in \mathrm{Lip}^*\alpha$ on each interval $[-\delta,\delta]$ with $\delta < 1$. This is obtained from the trigonometric case via the usual transformation $x = \cos\theta$. This type of an argument also shows that if $E_n(f) = o(n^{-r})$, then $f^{(r-1)}$ is smooth on $[-\delta,\delta]$ for each $\delta < 1$.

1.6. Approximation of Classes of Functions. The direct and inverse theorems of the preceding section combine to give characterizations of most classes of functions. For example, in the periodic case, the classes $\mathrm{Lip}^*\alpha$ are characterized by $f \in \mathrm{Lip}^*\alpha$ if and only if $E_n^*(f) = O(n^{-\alpha})$. In a similar way, we get a characterization for the classes $\mathrm{Lip}\,\alpha$, and $W^{(r)}(\alpha)$ when $0 < \alpha < 1$. When $\alpha = 1$ there is no characterization of these latter classes in terms of degree of approximation.

A somewhat less precise way of linking degree of approximation with smoothness is by the degree of approximation of classes. If $\mathcal{a}$ is a class of functions we define the error in approximating $\mathcal{a}$ by P_n (or T_n) by

$$E_n(\mathcal{a}) = \sup_{f\in\mathcal{a}} E_n(f)$$

$$E_n^*(\mathcal{a}) = \sup_{f\in\mathcal{a}} E_n^*(f)$$

The error in approximating α is controlled by the bad functions in α .

Theorem 1.8. Let $0 < \alpha \leq 1$, and $r \in N$ then there exist constants $C_1, C_2 > 0$ such that

$$C_1 n^{-r-\alpha} \leq E_n(W^{(r)}(\alpha,M,M_o, \ldots,M_r)) \leq C_2 n^{-r-\alpha} , \quad n \in N$$

The constants depend on the class but are independent of n . Similarly there are constants $C_1^*, C_2^* > 0$ such that

$$C_1^* n^{-r-\alpha} \leq E_n^* (W^{(r)}(\alpha,M,M_o, \ldots,M_r)) \leq C_2^* n^{-r-\alpha} , \quad n \in N$$

Also, if $0 < \alpha \leq 2$ then there are constants $C_3^*, C_4^* > 0$ such that

$$C_3^* n^{-\alpha} \leq E_n^*(\mathrm{Lip}^*(\alpha,M)) \leq C_4^* n^{-\alpha} , \quad n \in N$$

It is important to note that in each of the cases covered by Theorem 1.8, one can give an example of a single function f , in the appropriate class, such that $E_n(f)$ provides the lower estimate in Theorem 1.8.. For example, the function

$$f(t) = |t - (\tfrac{a+b}{2})|$$

is in $\mathrm{Lip}(1,1)$ and $E_n(f) \geq cn^{-1}$. Also $f(t) = |\sin t|$ is in $\mathrm{Lip}^*(1,1)$ and $E_n^*(f) \geq (4\pi n)^{-1}$.

1.7. Characterization of Classes of Functions in Terms of Fourier Coefficients

As we have mentioned, the class $W^{(r)}(1,M)$ is the same as the class of functions f for which

$$|f^{(r+1)}(x)| \le M \text{ a.e. in } [a,b] \tag{1.7.1.}$$

For the periodic case, (1.7.1) can also be expressed in terms of Fourier coefficients as

$$\sum_{-\infty}^{\infty} (ik)^{r+1} \hat{f}(k) e^{ikx} \in L_\infty(M) \tag{1.7.2.}$$

When r is odd, (1.7.2) can be written as

$$(-1)^{\frac{r+1}{2}} \sum_{0}^{\infty} k^{r+1} (a_k(f)\cos kx + b_k(f)\sin kx) \in L_\infty(M) \tag{1.7.3.}$$

When r is even then (1.7.3) still defines a class of functions but now we need the conjugate function to describe it.

Theorem 1.9. The class of functions f which satisfy

$$\sum_{0}^{\infty} k^r A_k(f,x) \sim g \in L_\infty \quad \text{and} \quad \| g \|_\infty \le M$$

is $W^{(r-1)}(1,M)$ when r is even and $\tilde{W}^{(r-1)}(1,M) = \{f: \widetilde{f^{(r-1)}} \in \mathrm{Lip}(1,M)\}$ when r is odd.

1.8. A Representation for Trigonometric Polynomials. It is common to represent trigonometric polynomials by interpolation formulae. We want to point out another representation of trigonometric polynomials which is not an interpolation formula (although it is very similar to one).

Let $x_k = 2k\pi(3n)^{-1}$, $k = 0,1, \ldots,3n-1$.

If T is a trigonometric polynomial of degree $\leq n$ then

$$T(x) = \frac{2}{3n} \sum_{k=0}^{3n-1} T(x_k)\, V_n(x-x_k) \tag{1.8.1.}$$

where V_n is the de la Vallée Poussin kernel

$$V_n(t) = \frac{\sin(\frac{3n}{2}t)\sin(\frac{nt}{2})}{2n\sin^2\frac{t}{2}} \tag{1.8.2.}$$

The reason (1.8.1) is not an interpolation formula in the usual sense is that (1.8.1) holds only for polynomials up to degree n while it has $3n$ nodes in the formula. However, the representation (1.8.1) is very good in the sense that if $x \in [-\pi,\pi]$, the terms in the sum (1.8.1) corresponding to points far away from x are very small. Indeed, if $x \neq x_k$ and $|x-x_k| \leq \pi$, then

$$|V_n(x-x_k)| \leq \frac{1}{2n}\left(\frac{(x-x_k)}{\pi}\right)^{-2} \tag{1.8.3.}$$

1.9. Orthogonal Polynomials and Quadrature Formulae. Let $w(x)$ be a non-negative function in $L_1[-1,1]$ with $\int_{-1}^{1} w(x)dx > 0$ and let (Q_n) be the sequence of orthogonal polynomials with respect to w . That is, $Q_n \in P_n$ and

$$\int_{-1}^{1} w(x)\, Q_i(x)\, Q_j(x)dx = \begin{cases} 0 & i \neq j \\ 1 & i = j \end{cases}$$

The zeros of Q_n are real, distinct, and all contained in $(-1,1)$. We label the zeros of Q_{2n} in increasing order as

$$-1 < x_{-n,2n} < \ldots < x_{-1,2n} < x_{1,2n} < \ldots < x_{n,2n} < 1 \qquad (1.9.1.)$$

Note that there is no $x_{o,2n}$. The zeros of Q_{2n+1} are denoted by

$$-1 < x_{-n,2n+1} < \ldots < x_{-1,2n+1} < x_{o,2n+1} < x_{1,2n+1} < \ldots < x_{n,2n+1} < 1 \qquad (1.9.2.)$$

When the weight w is an even function on $[-1,1]$, then the zeros of Q_{2n} and Q_{2n+1} are symmetric about the origin. That is

$$x_{k,2n} = -x_{-k,2n}, \quad x_{k,2n+1} = -x_{-k,2n+1} \text{ and } x_{o,2n+1} = 0 .$$

The zeros of Q_n are nodes of a quadrature formula which is exact for P_{2n-1}.

Theorem 1.10. (Gauss Quadrature Formula). There exist positive real numbers $A_k(2n)$, $k = -n, \ldots, -1, 1, \ldots, n$ such that for each $P \in P_{4n-1}$

$$\int_{-1}^{1} w(x)P(x)dx = \sum_{-n}^{n}{}' A_k(2n)P(x_{k,2n}) \qquad (1.9.3.)$$

Similarily, there exist positive real numbers $A_k(2n+1)$, $k = -n, \ldots, n$ such that for each $P \in P_{4n+1}$

$$\int_{-1}^{1} w(x)P(x)dx = \sum_{-n}^{n} A_k(2n+1)P(x_{k,2n+1}) \qquad (1.9.4.)$$

In (1.9.3), the notation $\sum'$ indicates that the term corresponding to $k = 0$ is omitted.

The Gauss quadrature formula has the property of highest possible precision for P_{2n-1}, that is, it is the only quadrature formula with $\leq n$ nodes which is precise for all $p \in P_{2n-1}$. The numbers $A_k(n)$ are known as the Cotes numbers for the quadrature formula. The numbers have a relation to the zeros of Q_n which is given in the following seperation theorem of Chebyshev, Markov, and Stieljes.

Theorem 1.11. *Let* $y_{k,2n}$ *be determined by* $y_{-n-1,2n} = -1$ *and*

$$A_k(2n) = \int_{-1}^{y_{k,2n}} w(t)dt - \int_{-1}^{y_{k-1,2n}} w(t)\, dt \quad .$$

Then $y_{n,2n} = 1$ *and*

$$x_{k,2n} < y_{k,2n} < x_{k+1,2n} \quad , \quad k = -n,\cdots,-1,1,\cdots,n-1 \quad .$$

More precisely, for $k = -n,\cdots,-1,1,\cdots,n$ *we have*

$$\int_{-1}^{y_{k-1,2n}} w(t)dt < \int_{-1}^{x_{k,2n}} w(t)dt < \int_{-1}^{y_{k,2n}} w(t)dt$$

$$= A_{-n}(2n) + \cdots + A_k(2n) \quad .$$

The analogous result holds for Q_{2n+1}.

The following Theorem of Bruns gives estimates on the size of the zeros of the Legendre polynomials. The Lengendre polynomials are the polynomials which are orthogonal with respect to the weight $w(x) = 1$ on $[-1,1]$.

Theorem 1.12. *Let* $x_{-n,2n} < \dots < x_{-1,2n} < x_{1,2n} < \dots < x_{n,2n}$ *be the zeros of the Legendre polynomial* P_{2n} *of degree* $2n$. *Then*

$$\cos\left(\frac{n+1-k}{2n+1}\pi\right) < x_{k,2n} < \cos\left(\frac{n+\frac{1}{2}-k}{2n}\pi\right) \qquad k = 1,2,\dots,n \ .$$

Similarily, for the zeros $x_{k,2n+1}$ *of* P_{2n+1} , *we have*

$$\cos\left(\frac{n+1-k}{2n+2}\pi\right) < x_{k,2n+1} < \cos\left(\frac{n+\frac{1}{2}-k}{2n+1}\pi\right) \qquad k = 1,2,\dots,n \ .$$

For the trigonometric case the quadrature formula with the highest degree of precision has a particularly simple form.

Theorem 1.13 *Let* x_o *be any point and* $x_k = x_o + \frac{2k\pi}{n}$ $k = 0,1, \dots,n-1$. *If* $T \in T_{n-1}$ *then*

$$\frac{1}{2\pi}\int_{-\pi}^{\pi} T(x)dx = \frac{1}{n}\sum_{k=0}^{n-1} T(x_k)$$

Thus we see that the nodes are equally spaced and the Cotes numbers are all equal.

1.10. Splines. For the material in this section, we refer the reader to any of the expository articles of I.J. Schoenberg (e.g. I.J. Schoenberg$_1$) . Let $m,n \in N$ and $a = x_o < x_1 < \dots < x_{n-1} < x_n = b$. Then $S_m(x_o, \dots,x_n)$ denotes the space of all splines of degree $m-1$ with knots at the points $x_1, \dots,x_{n-1}$ which have $m-2$ continuous derivatives. That is, $S \in S_m(x_o, \dots x_n)$ means that S is a polynomial of degree $\leq m-1$ on each interval $[x_i,x_{i+1}]$, $i = 0, \dots,n-1$ and S has $m-2$ continuous derivatives on $[a,b]$.

S_m is a linear subspace of $C[a,b]$ of dimension $m+n-1$. There is a canonical basis for S_m in terms of the so-called B-splines. Define $x_i = a$, $i \leq 0$ and $x_i = b$, $i \geq n$. For $f \in C[a,b]$, we let $f(x_i, \ldots, x_{i+m})$ denote the divided difference of f at $x_i, \ldots, x_{i+m}$. Then, $f(x_i, \ldots, x_{i+m})$ is the coefficient of x^m in the Taylor expansion (at 0) of the polynomial of degree m which interpolates f at the points $x_i, \ldots, x_{i+m}$. In the case that some of the points coalesce, say $x_k = x_{k+1} = \ldots = x_{k+r}$ then we require that f is r times differentiable. The polynomial is then understood to interpolate f and its' first r derivative at x_k.

Let $M(x;t) = m(t-x)_+^{m-1}$ where

$$(t-x)_+^{m-1} = \begin{cases} 0 & t \leq x \\ (t-x)^{m-1} & t \geq x \end{cases}$$

The B-splines are then defined for $-m+1 \leq i \leq n-1$ by

$$M_i(x) = M(x;x_i, \ldots, x_{i+m}) \tag{1.10.1}$$

where the divided difference is taken with respect to the variable t.

The B-splines have the following fundamental properties:

1.10.2. $M_i(x)$ vanishes outside $[x_i, x_{i+m}]$ and is non-negative on $[x_i, x_{i+m}]$.

1.10.3. $\int_a^b M_i(x)dx = 1$.

1.10.4. $\{M_i\}_{i=-m+1}^{n-1}$ is a basis for S_m.

CHAPTER 2

FUNDAMENTAL APPROXIMATION PROPERTIES OF POSITIVE LINEAR OPERATORS

2.1. Introduction. Let L be a linear operator mapping C^* into C^* (or $C[a,b]$ into $C[c,d]$, $[c,d] \subseteq [a,b]$). We say L is positive if for each non-negative function f in C^* the resulting function $L(f)$ is also non-negative. It follows that if $f \leq g$ then $L(f) \leq L(g)$. Many of the important linear methods of approximation are given by a sequence of positive linear operators. We cite some of the better known examples, which we will use throughout the text. Other examples will be introduced at appropriate times.

2.1.1 The Fejér Operators. Let $S_n(f)$ denote the nth partial sum of the Fourier series of $f \in C^*$. The Fejer sums of f are defined as

$$\sigma_n(f) = \frac{S_o(f) + \ldots + S_n(f)}{(n+1)}$$

For each $n \in N$, σ_n is a positive linear operator from C^* to T_n. This follows immediately from the kernel representation

$$\sigma_n(f) = f*F_n$$

with

$$F_n(t) = \frac{1}{2(n+1)} \left(\frac{\sin(n+1)(\frac{t}{2})}{\sin(\frac{t}{2})} \right)^2$$

2.1.2 The Jackson Operators. Let K_n be the Jackson kernel

$$K_n(t) = a_n \left(\frac{\sin(n+1)(\frac{t}{2})}{\sin(\frac{t}{2})} \right)^4$$

with a_n chosen so that

$$\frac{1}{\pi} \int_{-\pi}^{\pi} K_n(t)dt = 1$$

Then for $n \in N$, the operator J_n given by the convolution formula

$$J_n(f) = f*K_n$$

is a positive operator from C^* into T_{2n}. This again follows from the non-negativity of the kernel.

2.1.3 Trigonometric Convolution Operators. More generally we obtain a positive operator when we convolute with a non-negative Borel measure on $[-\pi,\pi)$. That is, for each non-negative measure $d\lambda$, the operator L given by

$$L(f) = f*d\lambda$$

is a positive linear operator from C^* to C^*. Of particular interest is when $d\lambda(t) = T(t)dt$, with T a non-negative trigonometric polynomial of degree n. Since in this case, for each $f \in C^*$, $L(f)$ is a trigonometric polynomial of degree $\leq n$.

2.1.4 The Bernstein Operators. For a function f in $C[0,1]$, the Bernstein polynomial of f of degree n is given by

$$B_n(f,x) = \sum_{k=0}^{n} \binom{n}{k} f(\tfrac{k}{n}) x^k (1-x)^{n-k} = \sum_{k=0}^{n} f(\tfrac{k}{n})\, p_{n,k}(x) .$$

The operator B_n is a positive linear operator from $C[0,1]$ into $C[0,1]$ and for each f, $B_n(f)$ is in P_n.

2.1.5 The Hermite-Fejer Operators. Let $T_n(x) = \cos(n \arccos x)$ be the Chebyshev polynomial of degree n. T_n has all its zeros in $(-1,1)$, at the points $x_{k,n} = \cos(\frac{2k-1}{2n}\pi)$, $k = 1,2,\ldots,n$. If $f \in C[0,1]$, let $H_n(f)$ be the polynomial of degree $\leq 2n-1$ which interpolates f at each $x_{k,n}$ and whose first derivative is zero at these points. That is,

$$H_n(f,x_{k,n}) = f(x_{k,n}) , \qquad k = 1,2,\ldots,n$$

and

$$H_n'(f,x_{k,n}) = 0 , \qquad k = 1,2,\ldots,n .$$

H_n is a positive operator mapping $C[-1,1]$ into $C[-1,1]$. The positivity of H_n can readily be seen from the representation

$$H_n(f,x) = \sum_{k=1}^{n} f(x_{k,n}) (1-xx_{k,n}) \left(\frac{T_n(x)}{n(x-x_{k,n})}\right)^2 \qquad -1 \leq x \leq 1$$

For each f in $C[-1,1]$, $H_n(f)$ is in P_{2n-1}.

2.1.6. Variation Diminishing Splines. Let $\Delta = \{0 = x_o < \ldots < x_{n-1} < x_n = 1\}$. As we have noted in Section 1.10, the B-splines M_i form a basis for $S_m(x_o, \ldots, x_n)$. We introduce the normalized B-splines

$$N_i(x) = \frac{x_{i+m} - x_i}{m} M_i(x) \qquad -m < i < n \quad .$$

Then, the function 1 has the simple representation

$$1 = \sum_{-m+1}^{n-1} N_i(x)$$

If we let

$$\xi_i = \frac{x_{i+1} + \cdots + x_{i+m-1}}{m-1}$$

and

$$\xi_i^{(2)} = \frac{x_{i+1}x_{i+2} + \ldots + x_{i+m-2}x_{i+m-1}}{\binom{m-1}{2}}$$

then we can represent x and x^2 as

$$x = \sum_{-m+1}^{n-1} \xi_i N_i(x)$$

$$x^2 = \sum_{-m+1}^{n-1} \xi_i^{(2)} N_i(x)$$

A simple developement for the above identities is given by M. Marsden[1].

I.J. Schoenberg has introduced the following operators called the variation diminishing splines. For $f \in C[0,1]$, we let

$$s_{\Delta,m}(f) = \sum_{-m+1}^{n-1} f(\xi_i)\, N_i(x)$$

Then $s_{\Delta,m}$ maps $C[0,1]$ into $C[0,1]$ and for each $f \in C[0,1]$, $s_{\Delta,m}(f) \in S_m(x_o, \ldots, x_n)$. Since each of the functions N_i are non-negative on $[0,1]$, $s_{\Delta,m}$ is a positive operator. These operators are a generalization of the Bernstein operators since when $\Delta = \{0,1\}$, $s_{\Delta,m+1}(f) = B_m(f)$, $f \in C[0,1]$.

2.1.7. Interpolation Operators. The Bernstein, Hermite-Fejer, and variation diminishing spline operators are examples of interpolation operators. The terminology is not meant in the sense that the resulting polynomial or spline interpolates the function at certain prescribed points. This is not the case for Bernstein polynomials. What is meant is that the values of the function at a certain finite number of points determines the result of operating on that function. If $x_1, x_2, \ldots, x_m$ are points in $[a,b]$ and h_k is a non-negative function in $C[c,d]$, $k = 1,2, \ldots, m$, then the operator

$$L(f,x) = \sum_{k=1}^{m} f(x_k)\, h_k(x)$$

is a positive linear operator from $C[a,b]$ into $C[c,d]$ which we call an interpolation operator.

2.1.8. The Landau Operators. For $n \in N$, let

$$c_n^{-1} = \int_{-1}^{1} (1-t^2)^n\, dt .$$

The Landau operator ℓ_n is given by the formula

$$\ell_n(f,x) = \int_{-\frac{1}{2}}^{\frac{1}{2}} f(t)\ c_n(1-(t-x)^2)^n\ dt\ .$$

It is a positive operator mapping $C[-\frac{1}{2},\frac{1}{2}]$ into $C[-\frac{1}{2},\frac{1}{2}]$, such that for each f in $C[-\frac{1}{2},\frac{1}{2}]$, $\ell_n(f)$ is a polynomial of degree $\leq 2n$.

2.1.9. Polynomial Convolution Operators. If P is any algebraic polynomial of degree n, which is non-negative on $[-1,1]$, then the operator L given by the convolution formula

$$L(f,x) = \int_{-\frac{1}{2}}^{\frac{1}{2}} f(t)\ P(t-x)\ dt$$

is a positive linear operator from $C[-\frac{1}{2},\frac{1}{2}]$ into $C[-\frac{1}{2},\frac{1}{2}]$, with the property that $L(f)$ is in P_n whenever f is in $C[-\frac{1}{2},\frac{1}{2}]$.

2.2. The Bohman-Korovkin Theorems. Our primary interest in these notes is to study the approximation properties of sequences of positive linear operators. If (L_n) is such a sequence, what are sufficient conditions to guarantee that $(L_n(f))$ converges uniformly to f for each continuous f? One of the important properties of positive operators is that it is enough to check a few simple functions. For example, we have the following two theorems of H. Bohman$_1$-P.P. Korovkin$_1$.

Theorem 2.1. Let (L_n) *be a sequence of positive linear operators mapping* $C[a,b]$ *into* $C[c,d]$, *with* $[c,d] \subseteq [a,b]$. *Suppose that for each* $i = 0,1,2,$

$$\|e_i - L_n(e_i)\| \, [c,d] \to 0$$

where $e_i(x) = x^i$. *Then for each* f *in* $C[a,b]$

$$\| f-L_n(f) \| \, [c,d] \to 0 \, .$$

Theorem 2.2. Let (L_n) *be a sequence of positive linear operators mapping* C^* *into* C^* . *Suppose that for* $i = 0,1,2,$

$$\| t_i - L_n(t_i) \| \to 0$$

where $t_o(x) = 1$, $t_1(x) = \sin x$, *and* $t_2(x) = \cos x$. *Then for each* f *in* C^*

$$\| f - L_n(f) \| \to 0 \, .$$

The proofs of these theorems follow from results in the next section.

2.3. *Quantatative Estimates.* It is possible to put the Bohman-Korovkin theorems in a quantitative form which estimates the rate of convergence of $(L_n(f))$ to f in terms of the rate of convergence for the test functions. The following theorem was given by R.Mamedov$_1$ in the case that $L_n(e_o) = e_o$ and in the general case by O. Shisha and B. Mond$_1$.

Theorem 2.3. Let (L_n) *be a sequence of positive linear operators mapping* $C[a,b]$ *into* $C[c,d]$ *with* $[c,d] \subseteq [a,b]$. *Define*

$$\alpha_n^2(x) = L_n((t-x)^2 \, , x) \qquad c \leq x \leq d \tag{2.3.1}$$

If $f \in C[a,b]$ *and* $x \in [c,d]$ *then*

$$|L_n(f,x) - f(x)| \le |f(x)|\ |e_o(x) - L_n(e_o,x)| + (L_n(e_o,x) + (L_n(e_o,x))^{\frac{1}{2}})\omega(f,\alpha_n(x))$$

(2.3.2)

If $f' \in C[a,b]$ *and* $x \in [c,d]$ *then*

$$|L_n(f,x) - f(x)| \le |f(x)|\ |e_o(x) - L_n(e_o,x)| + |f'(x)|\ |L_n((t-x),x)|$$

$$+ (1 + (L_n(e_o,x))^{\frac{1}{2}})\ \alpha_n(x)\omega(f',\alpha_n(x)) \qquad (2.3.3)$$

Remarks. This theorem recovers Theorem 2.1 since all the quantities on the right hand side of (2.3.2) tend to zero uniformly on $[c,d]$ when the hypothesis of Theorem 2.1 are satisfied. There are no claims that the constants in this theorem are best possible. In fact, as we shall see later, it is possible to improve the constants in some instances, although the question of best constants is not of primary concern to us. What is of concern is the rate of convergence to zero of our estimates and if they are the best possible results. This we will examine in detail in succeeding chapters(see also 2.10.3).

For the trigonometric case, the quantitative version of Theorem 2.2 is the following

Theorem 2.4. *Let* (L_n) *be a sequence of positive linear operators mapping* C^* *into* C^*. *Define*

$$\beta_n^2(x) = L_n(\sin^2(\tfrac{t-x}{2}),x), \quad -\pi \le x \le \pi \qquad (2.3.4)$$

<u>If</u> $f \in C^*$ and $x \in [-\pi,\pi]$ <u>then</u>

$$|L_n(f,x) - f(x)| \leq |f(x)|\ |t_o(x) - L_n(t_o,x)|$$

$$+ (L_n(t_o,x) + \pi(L_n(t_o,x))^{\frac{1}{2}})\,\omega(f,\beta_n(x)) \qquad (2.3.5)$$

<u>If</u> $f' \in C^*$ <u>and</u> $x \in [-\pi,\pi]$ <u>then</u>

$$|L_n(f,x) - f(x)| \leq |f(x)|\ |t_o(x) - L_n(t_o,x)| + |f'(x)| \left[|L_n(\sin(t-x),x)| \right.$$

$$\left. + \pi^2\beta_n^2(x) \right] + \pi^2(1 + (L_n(t_o,x))^{\frac{1}{2}})\ \beta_n(x)\ \omega(f',\beta_n(x)) \qquad (2.3.6)$$

<u>Proof.</u> The proofs of these two theorems are similar and we will only give the proof of Theorem 2.3.. First suppose that $f \in C[a,b]$ and $x \in [c,d]$. For any $t \in [a,b]$ and $\delta > \alpha_n(x)$, we use property 1.3.3 to find

$$|f(t) - f(x)| \leq \omega(f,|t-x|) \leq (1+\delta^{-1}|t-x|)\ \omega(f,\delta) \qquad (2.3.7)$$

Here if $\alpha_n(x) > 0$, we can take $\delta = \alpha_n(x)$. It follows from (2.3.7) that

$$|L_n(f,x) - f(x)| \leq L_n(|f(t) - f(x)|,x) + |L_n(f(x),x) - f(x)|$$

$$\leq \omega(f,\delta)\ (L_n(e_o,x) + \delta^{-1}\ L_n(|t-x|,x)) + |f(x)|\ |L_n(e_o,x) - e_o(x)|$$

From the Cauchy-Schwartz inequality for positive functionals we find

$$L_n(|t-x|,x) \le (L_n((t-x)^2,x))^{\frac{1}{2}}(L_n(e_o,x))^{\frac{1}{2}} = (L_n(e_o,x))^{\frac{1}{2}}\alpha_n(x)$$

Since $\delta > \alpha_n(x)$,

$$|L_n(f,x) - f(x)| \le |f(x)| \; |L_n(e_o,x) - e_o(x)| + (L_n(e_o,x) + (L_n(e_o,x))^{\frac{1}{2}})\omega(f,\delta) \qquad (2.3.8)\ .$$

The desired result (2.3.2) follows from (2.3.8) since δ is any number $> \alpha_n(x)$.

Now to the proof of (2.3.3) . Suppose $f' \in C[a,b]$ and write

$$f(t) = f(x) + f'(x)\ (t-x) + (f(t) - f(x) - f'(x)\ (t-x))$$

It follows that

$$|L_n(f,x) - f(x)| \le |f(x)| \; |L_n(e_o,x) - e_o(x)| + |f'(x)| \; |L_n(t-x,x)|$$

$$+ L_n(|f(t) - f(x) - f'(x)\ (t-x)|,x) \qquad (2.3.9)\ .$$

The first two terms are all right. The last term on the right hand side of (2.3.9) can be estimated by

$$|f(t) - f(x) - f'(x)\ (t-x)| \le |t-x|(1+\delta^{-1}|t-x|)\ \omega(f',\delta)$$

where δ is any number $> \alpha_n(x)$. Applying L_n, we find

$$L_n(|f(t) - f(x) - f'(x)(t-x)|,x) \le [L_n(|t-x|,x) + \delta^{-1}\alpha_n^2(x)]\omega(f',\delta)$$

$$\le [(L_n(e_o,x))^{\frac{1}{2}} \alpha_n(x) + \alpha_n^2(x)\ \delta^{-1}]\ \omega(f',\delta)$$

Since the only restriction on δ is that $\delta > \alpha_n(x)$, we find that

$$L_n(|f(t) - f(x) - f'(x)(t-x)|,x) \le ((L_n(e_o,x))^{\frac{1}{2}} + 1)\ \alpha_n(x)\ \omega(f',\alpha_n(x))\ .$$

This, together with (2.3.9) , gives (2.3.3) .

2.4. Convolution Operators. In special cases, the estimates in Theorems 2.3 and 2.4. become much simpler. For example, when (L_n) is a sequence of convolution operators on C^* we have the following corollary to Theorem 2.4 .

Corollary 2.1. *Let* (L_n) *be a sequence of positive convolution operators on* C^* . *That is, for each* $n \in N$

$$L_n(f) = f * d\lambda_n$$

where the measures $d\lambda_n$ *are even, non-negative, Borel measures on* $[-\pi,\pi]$ *with* $\frac{1}{\pi}\int_{-\pi}^{\pi} d\lambda_n(t) = 1$. *If*

$$\beta_n^2 = \frac{1}{\pi}\int_{-\pi}^{\pi} \sin^2\left(\frac{t}{2}\right) d\lambda_n(t) = \frac{1-\rho_{1,n}}{2}$$

then for each $f \in C^*$

$$||L_n(f) - f|| \leq (1+\pi)\omega(f,\beta_n) \leq (1+\pi)\omega(f,(1-\rho_{1,n})^{\frac{1}{2}}) \qquad (2.4.1).$$

When $f' \in C^*$

$$\| L_n(f)-f\| \leq \frac{\pi^2}{2} \|f'\| (1-\rho_{1,n}) + 2\pi^2 (1-\rho_{1,n})^{\frac{1}{2}} \omega(f',(1-\rho_{1,n})^{\frac{1}{2}}) \qquad (2.4.2)$$

Proof. We need only observe that $L_n(t_0) = t_0$, since the measures are normalized. Also, $L_n(\sin(t-x),x) = 0$, since the measures are even. Therefore, in the estimate (2.3.6), the first two terms on the right hand side of the inequality vanish. Note that $\beta_n(x) = \beta_n$ for all x.

A similar result holds in the algebraic case.

Corollary 2.2. For each $n \in N$, let L_n be a positive operator on $C[-\frac{1}{2},\frac{1}{2}]$ given by

$$L_n(f,x) = \int_{-\frac{1}{2}}^{\frac{1}{2}} f(t)\, d\lambda_n(t-x)$$

where each $d\lambda_n$ is a non-negative, even, Borel measure with unit mass on $[-1,1]$. Define

$$\alpha_n^2 = \int_{-1}^{1} t^2 \, d\lambda_n(t) .$$

If $\delta < \frac{1}{2}$, and $f \in C[-\frac{1}{2},\frac{1}{2}]$, then

$$|| L_n(f) - f||_\delta \leq (\tfrac{1}{2} - \delta)^{-2} \| f \|_\delta \alpha_n^2 + 2\omega(f,\alpha_n) \qquad (2.4.3)$$

If $f' \in C[-\frac{1}{2},\frac{1}{2}]$, then

$$\| L_n(f) - f\|_\delta \leq (\tfrac{1}{2} - \delta)^{-2}(\| f\|_\delta + \|f'\|_\delta)\, \alpha_n^2 + 2\alpha_n\, \omega(f', \alpha_n) \qquad (2.4.4).$$

Note: The subscript δ indicates that the norm is taken over $[-\delta,\delta]$.

Proof. To establish (2.4.3), we use (2.3.2) . Observe first that if $x \in [-\delta,\delta]$ then

$$L_n((t-x)^2,x) = \int_{-\frac{1}{2}}^{\frac{1}{2}} (t-x)^2\, d\lambda_n(t-x) \leq \int_{-1}^{1} t^2\, d\lambda_n(t) = \alpha_n^2 . \qquad (2.4.5)$$

Also,

$$L_n(e_o,x) = \int_{-\frac{1}{2}-x}^{\frac{1}{2}-x} d\lambda_n(t) \leq \int_{-1}^{1} d\lambda_n(t) = 1 \qquad (2.4.6)$$

We estimate the error in approximating e_o by

$$e_o(x) - L_n(e_o,x) = \int_{-1}^{1} d\lambda_n(t) - \int_{-\frac{1}{2}}^{\frac{1}{2}} d\lambda_n(t-x) = \int_{-1}^{-\frac{1}{2}-x} d\lambda_n(t) + \int_{\frac{1}{2}-x}^{1} d\lambda_n(t)$$

$$\leq 2\int_{\frac{1}{2}-\delta}^{1} d\lambda_n(t) \leq 2(\tfrac{1}{2} - \delta)^{-2} \int_{\frac{1}{2}-\delta}^{1} t^2\, d\lambda_n(t) \leq (\tfrac{1}{2} - \delta)^{-2}\, \alpha_n^2 \qquad (2.4.7)$$

Using (2.4.5), (2.4.6) and (2.4.7) in (2.3.2) gives the desired result. Now for (2.4.4), we need to estimate $|L_n((t-x),x)|$. Since $d\lambda_n$ is an even measure, we have

$$|L_n((t-x),x)| = |\int_{-\frac{1}{2}-x}^{\frac{1}{2}-x} t\, d\lambda_n(t)| \leq 2\int_{\frac{1}{2}-\delta}^{1} |t|\, d\lambda_n(t)$$

$$\leq 2(\tfrac{1}{2}-\delta)^{-1}\int_{\frac{1}{2}-\delta}^{1} t^2\, d\lambda_n(t) \leq (\tfrac{1}{2}-\delta)^{-1}\alpha_n^2 \qquad (2.4.8)$$

We now use (2.4.5), (2.4.6), (2.4.7) and (2.4.8) in (2.3.3) to find

$$\|L_n(f) - f\|_\delta \leq \|f\|_\delta (\tfrac{1}{2}-\delta)^{-2}\alpha_n^2 + \|f'\|_\delta (\tfrac{1}{2}-\delta)^{-1}\alpha_n^2 + 2\alpha_n\,\omega(f',\alpha_n)$$

$$\leq (\tfrac{1}{2}-\delta)^{-2}(\|f\|_\delta + \|f'\|_\delta)\,\alpha_n^2 + 2\alpha_n\,\omega(f',\alpha_n)$$

which is (2.4.4).

2.5. Estimates for the Fejér, Jackson and Landau operators

2.5.1. The Fejér Operators. For the Fejér operators, it is easy to evaluate the first Fourier coefficients. We have

$$1-\rho_{1,n} = \frac{2}{\pi}\int_{-\pi}^{\pi} \sin^2(\tfrac{t}{2})F_n(t)\,dt = \frac{1}{\pi}\int_{-\pi}^{\pi}(n+1)^{-1}\ \sin^2(n+1)(\tfrac{t}{2})dt = (n+1)^{-1}$$

Thus, from Corollary 2.1, if $f \in C^*$

$$\|f-\sigma_n(f)\| \leq (1+\pi)\,\omega(f,n^{-\frac{1}{2}})$$

and if f is continuously differentiable on $[-\pi,\pi]$, then

$$\|f-\sigma_n(f)\| \leq 2\pi^2 n^{-\frac{1}{2}}\,\omega(f',n^{-\frac{1}{2}}) + \frac{\pi^2}{2}\|f'\|\,n^{-1}$$

2.5.2. Jackson operators. We first evaluate

$$a_{n-1}^{-1} = \frac{1}{\pi} \int_{-\pi}^{\pi} \left(\frac{\sin(\frac{nt}{2})}{\sin(\frac{t}{2})} \right)^4 dt$$

We have for the Fejér kernel

$$F_{n-1}(t) = (2n)^{-1} \left(\frac{\sin(\frac{nt}{2})}{\sin(\frac{t}{2})} \right)^2 = \frac{1}{2} + \sum_{k=1}^{n-1} (1 - \frac{k}{n}) \cos kt$$

This gives that

$$a_{n-1}^{-1} = 4n^2 \pi^{-1} \int_{-\pi}^{\pi} (\frac{1}{2} + \sum_{k=1}^{n-1} (1 - \frac{k}{n}) \cos kt)^2 dt$$

$$= 4n^2 (\frac{1}{2} + \sum_{k=1}^{n-1} (1 - \frac{k}{n})^2) = 2n^2 + 4 \sum_{k=1}^{n-1} k^2 = \frac{4n^3+2n}{3}$$

Since the coefficient of $\cos t$ in the trigonometric expansion of $(\sin(\frac{t}{2}))^{-4} (\sin(\frac{nt}{2}))^4$ is

$$4n^2 \sum_{k=1}^{n-1} (1 - \frac{k}{n})(1 - \frac{k-1}{n})$$

we have

$$\frac{1}{\pi}\int_{-\pi}^{\pi}\sin^2(\tfrac{t}{2})K_{n-1}(t)\,dt = \frac{a_{n-1}}{2\pi}\int_{-\pi}^{\pi}(1-\cos t)\left(\frac{\sin(\frac{nt}{2})}{\sin(\frac{t}{2})}\right)^4 dt$$

$$= a_{n-1}\left[\frac{1}{2}a_{n-1}^{-1} - 2n^2\sum_{k=1}^{n-1}(1-\frac{k}{n})(1-\frac{k-1}{n})\right]$$

$$= a_{n-1}\left[2n^2(\frac{1}{2}+\sum_{k=1}^{n-1}(1-\frac{k}{n})^2) - 2n^2\sum_{k=1}^{n-1}(1-\frac{k}{n})(1-\frac{k-1}{n})\right]$$

$$= a_{n-1}\ 2n^2(\frac{1}{2} - n^{-1}\sum_{k=1}^{n-1}(1=\frac{k}{n})) = na_{n-1} = \frac{3}{4}n^{-2} + O(n^{-3})\ .$$

In otherwords,

$$1-\rho_{1,n} = 3(2n^2+4n+3)^{-1} = \frac{3}{2}n^{-2} + O(n^{-3})$$

If we use this estimate in Corollary 2.1, we find that if $f \in C^*$

$$\| f-J_n(f) \| \leq (1+\pi)\ \omega(f,n^{-1})$$

and if $f' \in C^*$ then

$$\| f-J_n(f) \| \leq 6\pi^2\ n^{-1}\ \omega(f',n^{-1}) + \pi^2\|f'\|\ n^{-2}\ .$$

These are the classical estimates of Jackson_1 which serve as the basis for the the direct theorems on approximation (see Section 1.5) .

2.5.3. Landau Operators. We first recall the following well known identity.

$$\int_0^1 (1-t)^{p-1}\, t^{q-1}\, dt = \frac{\Gamma(p)\Gamma(q)}{\Gamma(p+q)} .$$

The evaluation of c_n^{-1} is easy.

$$c_n^{-1} = \int_{-1}^1 (1-t^2)^n\, dt = 2\int_0^1 (1-t^2)^n\, dt$$

$$= \int_0^1 (1-t)^n\, t^{-\frac{1}{2}} dt = \frac{\Gamma(n+1)\Gamma(\frac{1}{2})}{\Gamma(n+\frac{3}{2})} .$$

This gives for the second moments.

$$c_n \int_{-1}^1 t^2(1-t^2)^n\, dt = 2c_n \int_0^1 t^2(1-t^2)^n dt$$

$$= c_n \int_0^1 t^{\frac{1}{2}}\, (1-t)^n dt = \frac{\Gamma(n+\frac{3}{2})}{\Gamma(n+1)\Gamma(\frac{1}{2})} \cdot \frac{\Gamma(n+1)\Gamma(\frac{3}{2})}{\Gamma(n+\frac{5}{2})}$$

$$= (2n+3)^{-1}$$

Using this in Corollary 2.2 gives the following results of R. Mamedov$_2$. If $\delta < \frac{1}{2}$, and $f \in C[-\frac{1}{2},\frac{1}{2}]$, then

$$\| f-\ell_n(f)\| \leq \frac{1}{2}\, \| f\|_\delta\, (\tfrac{1}{2} - \delta)^{-2}\, n^{-1} + 2\omega(f,n^{-\frac{1}{2}})$$

and if $f' \in C[-\frac{1}{2},\frac{1}{2}]$, then

$$\| f-\ell_n(f) \| \leq \frac{1}{2} (\| f \|_\delta + \| f' \|_\delta)(\tfrac{1}{2} - \delta)^{-2} n^{-1} + 2n^{-\frac{1}{2}} \omega(f', n^{-\frac{1}{2}}) .$$

We also want to record for future use the following equality for the fourth moments

$$c_n \int_{-1}^{1} t^4 (1-t^2)^n \, dt = 2c_n \int_0^1 t^4 (1-t^2)^n dt = c_n \int_0^1 t^{3/2} (1-t)^n \, dt =$$

$$= \frac{\Gamma(n + \frac{3}{2}) \Gamma(\frac{5}{2})}{\Gamma(\frac{1}{2}) \Gamma(n + \frac{7}{2})} = 3(2n+5)^{-1} (2n+3)^{-1}$$

<u>2.6. Finer Estimates for Twice Differentiable Functions.</u> Although we have not concerned ourselves much with the constants appearing in our estimates, there is one case where it is easy to obtain the best constant. This is when we are dealing with functions whose first derivative is in Lip(1,M) or equivalently $|f''| \leq M$, a.e..

<u>Theorem. 2.5.</u> *Let* (L_n) *be a sequence of positive linear operators mapping* $C[a,b]$ *into* $C[c,d]$, *with* $[c,d] \subseteq [a,b]$, *and let*

$$\alpha_n^2 (x) = \frac{1}{2} L_n((t-x)^2, x) \qquad c \leq x \leq d \tag{2.6.1}$$

If for $i = 0,1$ *and* $x_o \in [c,d]$, *we have*

$$e_i(x_o) - L_n(e_i, x_o) = o(\alpha_n^2(x_o)) \tag{2.6.2}$$

then when $f' \in Lip(1,M)$ on $[a,b]$

$$|f(x_o) - L_n(f,x_o)| \le M\alpha_n^2(x_o) + o(\alpha_n^2(x_o)) \qquad (2.6.3)$$

Remarks. If $L_n(e_i,x_o) = e_i(x_o)$, $i = 0,1$, then the $o(\alpha_n^2(x_o))$ term in (2.6.3) can be dropped. Since the function $\frac{1}{2}M(t-x_o)^2$ has a second derivative $\le M$ everywhere on $[a,b]$, we see that (2.6.3) is a sharp estimate. If the estimate (2.6.2) is uniform in $x_o \in [c,d]$ then likewise the "o" term in (2.6.3) is uniform in $x_o \in [c,d]$.

Proof. If $t \in [a,b]$, then

$$f(t) = f(x_o) + f'(x_o)(t-x_o) + \int_{x_o}^{t}\int_{x_o}^{s} f''(u)\,du\,ds \qquad (2.6.4)$$

Since,

$$\left| \int_{x_o}^{t}\int_{x_o}^{s} f''(u)\,du\,ds \right| \le \frac{1}{2} M(t-x_o)^2$$

if we apply L_n to the right hand side of (2.6.4), we find

$$|L_n(f,x_o) - f(x_o)| \le |f(x_o) - L_n(f(x_o),x_o)| + |f'(x_o)||L_n((t-x_o),x_o|$$

$$+ \frac{1}{2} M\, L_n((t-x_o)^2,x_o) \le M\alpha_n^2(x_o) + o(\alpha_n^2(x_o)) .$$

Here, the first two terms were replaced by $o(\alpha_n^2(x_o))$ because of (2.6.2).

<u>2.7. Estimates for Bernstein Operators</u>. To estimate $e_i(x) - B_n(e_i,x)$ we will need to evaluate the sums

$$b_{n,r}(x) = \sum_{k=0}^{n} (k-nx)^r \, p_{n,k}(x) \qquad 0 \le x \le 1 \qquad (2.7.1)$$

when $r = 0,1,2$.

For $r = 0$, it is clear that $b_{n,0}(x) = 1$, for $0 \le x \le 1$. This gives that $B_n(e_0) = e_0$ on $[0,1]$. When $r = 1$,

$$b_{n,1}(x) = \sum_{k=0}^{n} (k-nx)p_{n,k}(x) = \sum_{k=1}^{n} k\binom{n}{k} x^k(1-x)^{n-k} - nx =$$

$$= nx \sum_{k=1}^{n} \binom{n-1}{k-1} x^{k-1}(1-x)^{n-1-(k-1)} - nx = 0 \qquad (2.7.2)$$

This gives that $B_n(e_1) = e_1$ on $[0,1]$.

To evaluate $b_{n,2}$, we will verify the recurrence formula

$$b_{n,r+1}(x) = x(1-x)(b'_{n,r}(x) + nrb_{n,r-1}(x)) \qquad \text{for } r \ge 1 \qquad (2.7.3)$$

For any $0 \le k \le n$,

$$p'_{n,k}(x) = x^{-1}(1-x)^{-1}(k-nx)\, p_{n,k}(x)$$

Therefore,

$$x(1-x)(b'_{n,r}(x) + nrb_{n,r-1}(x)) = x(1-x) \sum_{k=0}^{n} \Big[(k-nx)^r p'_{n,k}(x) - nr(k-nx)^{r-1} p_{n,k}(x) +$$

$$+ nr(k-nx)^{r-1}p_{n,k}(x)\Big] = x(1-x) \sum_{k=0}^{n} x^{-1}(1-x)^{-1}(k-nx)^{r+1}p_{n,k}(x) = b_{n,r+1}(x)$$

which is (2.7.3). This recurrence formula together with our evaluation of $b_{n,o}$ and $b_{n,1}$ show that $b_{n,2}(x) = nx(1-x)$. From this it follows that

$$B_n((t-x)^2,x) = \sum_{k=0}^{n} \left(\frac{k}{n} - x\right)^2 p_{n,k}(x) = x(1-x)n^{-1} .$$

Using Theorem 2.3, we find that if $f \in C[0,1]$,

$$|f(x) - B_n(f,x)| \leq 2\omega(f,x^{\frac{1}{2}}(1-x)^{\frac{1}{2}}n^{-\frac{1}{2}}), \quad 0 \leq x \leq 1 \qquad (2.7.4)$$

and if $f' \in C[0,1]$

$$|f(x) - B_n(f,x)| \leq 2x^{\frac{1}{2}}(1-x)^{\frac{1}{2}}n^{-\frac{1}{2}} \omega(f',x^{\frac{1}{2}}(1-x)^{\frac{1}{2}}n^{-\frac{1}{2}}) \; 0 \leq x \leq 1 \qquad (2.7.5)$$

These estimates are due to T. Popoviciu_1.

When we use Theorem 2.5, we see that if $f' \in \mathrm{Lip}(1,M)$ then

$$|f(x) - B_n(f,x)| \leq \frac{1}{2} M \, x(1-x)n^{-1} \qquad 0 \leq x \leq 1 \qquad (2.7.6)$$

2.8. <u>Estimates for the Hermite-Fejér Operators</u>. We have mentioned before that $H_n(f)$ is the polynomial of degree $2n-1$ which interpolates the value of f at the zeros of the Chebyshev polynomial and has a derivative equal to zero at these points. From this, it follows that

$$H_n(e_o,x) = \sum_{k=1}^{n} (1-xx_{k,n})\left(\frac{T_n(x)}{n(x-x_{k,n})}\right)^2 = 1 \qquad -1 \leq x \leq 1 \qquad (2.8.1).$$

We could also show this directly.

For e_1, we have

$$H_n(e_1,x) - e_1(x) = \sum_{k=1}^{n} (x_{k,n}-x)(1-xx_{k,n})\left(\frac{T_n(x)}{n(x-x_{k,n})}\right)^2$$

$$= - \sum_{k=1}^{n} [(1-x^2)(x-x_{k,n}) + x(x-x_{k,n})^2]\left(\frac{T_n(x)}{n(x-x_{k,n})}\right)^2$$

$$= - n^{-2}(1-x^2)\ T_n^2(x) \sum_{k=1}^{n} (x-x_{k,n})^{-1} - n^{-1}\ x\ T_n^2(x)$$

$$= -n^{-2}(1-x^2)\ T_n(x)\ T_n'(x) - n^{-1}\ x\ T_n^2(x) \tag{2.8.2}$$

Now,

$$|(1-x^2)\ T_n'\ (x)| = |(1-x^2)^{\frac{1}{2}}\ n \sin n(\arccos x)| \leq n$$

Therefore, (2.8.2) can be estimated by

$$|H_n(t-x),x)| \leq 2n^{-1}|T_n(x)| \tag{2.8.3}$$

The estimate of $H_n((t-x)^2,x)$ is easy since

$$H_n((t-x)^2,x) = \sum_{k=1}^{n} (x_{k,n}-x)^2(1-xx_{k,n}) \left(\frac{T_n(x)}{n(x-x_{k,n})}\right)^2$$

$$= n^{-2}\ T_n^2(x) \sum_{k=1}^{n} (1-xx_{k,n}) = n^{-1}\ T_n^2(x) \tag{2.8.4}$$

since the points $x_{k,n}$ are symmetric about the origin.

When we use (2.8.1), (2.8.3), and (2.8.4) in Theorem 2.3, we find that if $f \in C[-1,1]$, then

$$|f(x) - H_n(f,x)| \le 2\omega(f, n^{-\frac{1}{2}}\, |T_n(x)|) \qquad (2.8.5)$$

and if $f' \in C[-1,1]$ then

$$|f(x) - H_n(f,x)| \le 2|T_n(x)|(n^{-1}|f'(x)| + n^{-\frac{1}{2}}\,\omega(f', n^{-\frac{1}{2}}|T_n(x)|) \qquad (2.8.6)$$

These estimates were given by T. Popoviciu$_2$ and independently by O. Shisha, C. Sternin and M. Fekete$_1$. For improvements on these estimates see Theorem 7.11.

2.9. Estimates for Variation Diminishing Splines. The variation diminishing splines preserve linear functions and so we need only estimate $s_{\Delta,m}((t-x)^2, x)$. We have following (2.1.6) that

$$s_{\Delta,m}((t-x)^2, x) = s_{\Delta,m}(t^2, x) - x^2 = \sum_{-m+1}^{n-1} (\xi_i^2 - \xi_i^{(2)})\, N_i(x) \qquad (2.9.1)$$

We can find a closed form for $\xi_i^2 - \xi_i^{(2)}$ as follows

$$\xi_i^2 - \xi_i^{(2)} = (m-1)^{-2}\left(\sum_{r=1}^{m-1} x_{i+r}\right)^2 - 2(m-1)^{-1}(m-2)^{-1} \sum_{1\le r<s\le m-1} x_{i+r}x_{i+s}$$

$$= (m-1)^{-2}\left(2\sum_{1\le r<s\le m-1} x_{i+r}x_{i+s} + \sum_{r=1}^{m-1} x_{i+r}^2\right) - 2(m-1)^{-1}(m-2)^{-1}\sum_{1\le r<s\le m-1} x_{i+r}x_{i+s}$$

$$= (m-1)^{-2}(m-2)^{-1}\left(-2\sum_{1\le r<s\le m-1} x_{i+r}x_{i+s} + (m-2)\sum_{r=1}^{m-1} x_{i+r}^2\right)$$

$$= (m-1)^{-2}(m-2)^{-1}\sum_{1\le r<s\le m-1} (x_{i+r} + x_{i+s})^2 \qquad (2.9.2)$$

Because of (2.9.2) , we define

$$\alpha^2_{\Delta,m} = (m-1)^{-1} \max_{-m+1 \le i \le n} (x_{i+m-1} - x_i)^2$$

so that

$$s_{\Delta,m}((t-x)^2,x) = \sum_{-m+1}^{n-1} (\xi_i^2 - \xi_i^{(2)}) N_i(x) \le \alpha^2_{\Delta,m} \sum_{-m+1}^{n-1} N_i(x) = \alpha^2_{\Delta,m} \quad (2.9.3)$$

It should be noted that in the case that the degree m is bounded then $\alpha^2_{\Delta,m}$ is asymptotically equivalent to $||\Delta||^2$ where

$$\| \Delta \| = \max_{0 \le i \le n-1} |x_{i+1} - x_i| \quad (2.9.4)$$

If we use Theorem 2.3 , we get the following estimate for approximation by variation diminishing splines. If $f \in C[0,1]$ then

$$\|f-s_{\Delta,m}(f)\| \le 2\omega(f,\alpha_{\Delta,m}) \quad (2.9.5)$$

If $f' \in C[0,1]$ then

$$\| f-s_{\Delta,m}(f) \| \le 2\alpha_{\Delta,m}\, \omega(f',\alpha_{\Delta,m}) \quad (2.9.6)$$

2.10. Notes

2.10.1. There are several directions in which the Bohman-Korovkin Theorems can be extended. For instance, the test functions $\{1,t,t^2\}$ can be replaced by other sets of test functions. Such a permissible set of test functions is commonly called a Korovkin system. A characterization of Korovkin systems can be given in terms of the Choquet boundary (see Yu. Šaškin$_1$ or D. Wulbert$_1$). Namely, a necessary and sufficient condition that $\{u_1, \ldots, u_k\}$ be a Korovkin system for $C[a,b]$ is that the Choquet boundary of $sp(u_1, \ldots, u_k)$ is all of $[a,b]$. There is no known way of obtaining quantitative estimates (analogous to Theorems 2.3 and 2.4) for arbitrary Korovkin systems. G. Freud$_2$ has obtained quantitative estimates for the case that $\{u_1, \ldots, u_k\}$ is an extended complete Chebyshev system (see also E. Censor$_1$).

2.10.2. Korovkin systems have also been studied in the L_p-spaces, primarily by M. Müller$_1$. Among other things, it is known that $\{1,t,t^2\}$ and $\{1,\sin t,\cos t\}$ are Korovkin systems for L_p although there is no known characterization of Korovkin systems in these spaces. An appropriate boundary has not yet been given for the L_p-spaces. Quantitative theroems, even for $\{1,t,t^2\}$ are not known in L_p except for special cases such as convolution operators. Quantitative theorems are known for the space $C(-\infty,\infty)$ (see Müller and Walk$_1$ or Ditzian$_1$).

2.10.3. As we have pointed out, the estimate in Theorem 2.5 is sharp for the class of functions $W^{(1)}(1,M)$. It is also possible to give asymptotically sharp estimates for other classes of functions. These estimates need not be the same as those given in this chapter. For example, the asymptotically sharp estimate for the class $\mathrm{Lip}(\alpha,M)$ is of the order $L_n(|t-x|^\alpha,x)$, while in this chapter our estimate is $(L_n((t-x)^2,x))^{\frac{\alpha}{2}}$, which is never better and could be worse. An instructive example is the Fejér operators, for which $\sigma_n(|t-x|^\alpha,x) \sim n^{-\alpha}$, when $0 < \alpha < 1$. This gives the order of approximation $n^{-\alpha}$ for the class $\mathrm{Lip}(\alpha,M)$. Thus far, we have only obtained the poorer estimate $n^{-\frac{\alpha}{2}}$. A more detailed discussion of best estimates for classes is given in Chapter 7 in the context of the degree of approximation of classes of functions.

CHAPTER 3

SATURATION OF CONVOLUTION OPERATORS

3.1. Introduction. Until now, we have concentrated on giving estimates for the degree of approximation by positive operators. There are three common ways of measuring the precision of such estimates. They are saturation, approximation of classes of functions, and inverse theorems. We begin in this chapter by discussing saturation for the important case of convolution on $C^*[-\pi,\pi]$.

Let (L_n) be a sequence of positive convolution operators. That is, for each $n \in N$, we assume that L_n has the form

$$L_n(f,x) = (f*d\mu_n)(x) = \frac{1}{\pi}\int_{-\pi}^{\pi} f(x+t)d\mu_n(t) \qquad (3.1.1)$$

where $d\mu_n$ is a non-negative, even, Borel measure on $[-\pi,\pi)$ with

$$\frac{1}{\pi}\int_{-\pi}^{\pi} d\mu_n(t) = 1 .$$

To motivate the concept of saturation, let us recall that the estimates in Chapter 2 (the one applicable here is Corollary 2.1) were of the form

$$\| f-L_n(f)\| \leq C\omega(f,\beta_n) \qquad (3.1.2)$$

for continuous functions; where $\beta_n^2 = (1-\rho_{1,n}) = \frac{2}{\pi}\int_{-\pi}^{\pi} \sin^2\frac{t}{2}\, d\mu_n(t)$. When f is continuously differentiable the estimate took the form

$$\| f-L_n(f) \| < C\beta_n \omega(f',\beta_n) \tag{3.1.3}$$

In particular when $f' \in \text{Lip } 1$

$$\| f-L_n(f) \| \le C\beta_n^2 \tag{3.1.4}$$

Suppose f is a function which has two continuous derivatives. Is it then possible to improve the estimate (3.1.4)? Generally, it is not. For example, the function $f_o(t) = \cos t$ is infinitely differentiable and yet

$$|f_o(0) - L_n(f_o,0)| = 1 - \frac{1}{\pi} \int_{-\pi}^{\pi} \cos t \, d\mu_n(t) = 1-\rho_{1,n} = \beta_n^2 .$$

Thus, positive convolution operators are limited in their efficiency of approximation, at least in the sense that higher smoothness of the function f being approximated generally does not guarantee a better degree of approximation.

It still may happen that there are functions which are approximated with order $o(\beta_n^2)$. Of course, this is the case for constant functions. When it only happens for constant functions, we will say (L_n) is saturated.

<u>Definition.</u> <u>Let (L_n) be a sequence of operators from C^* into C^* and (ϕ_n) a sequence of positive real numbers which converge to 0. We say that (L_n) is saturated with order (ϕ_n) if the following two conditions are satisfied:</u>

<u>3.1.5. If $f \in C^*$, then</u>

$$\lim_{n\to\infty} \frac{\|f-L_n(f)\|}{\phi_n} = 0$$

if, and only if, f *is a constant.*

3.1.6. There is a non-constant function $f_o \in C^*$ *for which*

$$\| f_o - L_n(f_o) \| = O(\phi_n) .$$

The problem of saturation is to determine when (L_n) is saturated, and if so, the *saturation order* (ϕ_n) and the *saturation class* $S(L_n)$, which consists of all those functions f_o which satisfy (3.1.6). A saturation order (ϕ_n) of (L_n) (if it exists) is not unique. For example, if (ϕ'_n) is any positive sequence for which there are constants $0 < C_1 < C_2 < +\infty$ such that

$$C_1 \leq \frac{\phi'_n}{\phi_n} \leq C_2 , \qquad n \in N \tag{3.1.7}$$

then (ϕ'_n) is also a saturation order. However, it is easy to see that any two saturation orders of (L_n) are related as in (3.1.7).

For suppose (L_n) is saturated with order (ϕ_n) and also with order (ϕ'_n). If

$$\underline{\lim}_{n\to\infty} \frac{\phi'_n}{\phi_n} = 0$$

then there is a subsequence (n_k) for which $\phi'_{n_k} = o(\phi_{n_k})$. If f_o is the non-constant function guaranteed by (3.1.6) for (ϕ'_n), then

$$\|f_o - L_{n_k}(f_o)\| = O(\phi'_{n_k}) = o(\phi_{n_k})$$

which contradicts (3.1.5) for (ϕ_n). Therefore, there is a constant $C_1 > 0$ such that for $n \in N$, $\phi'_n\phi_n^{-1} \geq C_1$. To prove the existence of C_2, we merely reverse the roles of (ϕ'_n) and (ϕ_n).

We shall use the notation $(\phi_n) \sim (\phi'_n)$ to mean that the two sequences are related as in (3.1.7). The relationship "$\sim$" is then an equivalence relation on sequences and all saturation orders for (L_n) lie in the same equivalence class. It is important to note that the saturation class $S(L_n)$ is well defined and does not depend on which representative we choose as the saturation order. Thus in the sequel, when we determine the saturation class, we will work with whatever saturation order is most convenient.

The condition (3.1.5) is not the one usually found in the literature. Generally, it is assumed only that $\|f - L_n(f)\| = o(\phi_n)$ implies that f is constant. However, as we shall see in Section 3.10, under this definition, it is possible to have two saturation orders (ϕ_n) and (ϕ'_n) that are not related as in (3.1.7). More importantly, the saturation classes for these two orders may be different. All the classical saturation theorems come out the same under our slightly more restrictive definition.

<u>3.2. Two Simple Examples</u>. The discussion in 3.1 may have suggested that the saturation of (L_n) is always determined by the first Fourier-Stieljes coefficients of $d\mu_n$. The situation is not that simple, and we will now give two examples which will indicate more clearly the role of Fourier coefficients in determining the saturation order.

<u>Example 1.</u> Consider the sequence of positive operators (A_n) given by

$$A_n(f) = f * d\alpha_n \ , \quad n \in N$$

where

$$d\alpha_n = \frac{\pi}{2}\,(d\rho_{-n^{-1}} + d\rho_{n^{-1}})$$

We use the notation $d\rho_{x_o}$ to denote the Dirac measure at x_o, (i.e. for $f \in C^*$, $\int_{-\pi}^{\pi} f(t)\,d\rho_{x_o}(t) = f(x_o)$).

We now examine the real Fourier-Stieljes coefficients $\rho_{k,n}$ of $d\alpha_n$, for the purpose of determining the saturation order of (A_n). For $k, n \in N$

$$1-\rho_{k,n} = 1-\cos\frac{k}{n} = 2\sin^2\frac{k}{2n}$$

Thus, for $k \in N$

$$\lim_{n\to\infty} n^2(1-\rho_{k,n}) = \frac{k^2}{2} \ .$$

Now we can easily show (A_n) is saturated with order (n^{-2}). To establish condition (3.1.5), we need only show that if for some subsequence (n_j)

$$\| f-A_{n_j}(f) \| = o(n_j^{-2})$$

then f is constant. So, suppose $f \in C^*$ and

$$\| f-A_{n_j}(f) \| = o(n_j^{-2})$$

Then,

$$o(n_j^{-2}) = \frac{1}{2\pi}\int_{-\pi}^{\pi} (f-f*d\alpha_{n_j})e^{-ikt}dt =$$

$$\frac{1}{2\pi}\int_{-\pi}^{\pi} f(x)e^{-ikx}dx - \frac{1}{2\pi}\int_{-\pi}^{\pi} e^{-ikx}\left[\frac{1}{\pi}\int_{-\pi}^{\pi} f(x+t)\,d\alpha_{n_j}(t)\right]$$

$$= \hat{f}(k) - 2\check{\alpha}_{n_j}(k)\,\hat{f}(k) = \hat{f}(k)(1-\rho_{k,n_j})$$

Therefore

$$\frac{k^2\hat{f}(k)}{2} = \lim_{j\to\infty} n_j^2\,\hat{f}(k)(1-\rho_{k,n_j}) = 0$$

and so, $\hat{f}(k) = 0$, $k = \pm 1, \pm 2, \ldots$. This shows that f is constant.

We verify condition (3.1.6) by considering the function $f_o(t) = \cos t$ Using the fact that the measure $d\alpha_n$ is even, we find

$$\| f_o-A_n(f_o) \| = \left\| \frac{1}{\pi}\int_{-\pi}^{\pi} [\cos x - \cos(x+t)]\,d\alpha_n(t) \right\|$$

$$= \left\| \frac{1}{\pi}\int_{-\pi}^{\pi} \cos x(1-\cos t)\,d\alpha_n(t) + \frac{1}{\pi}\int_{-\pi}^{\pi} \sin x \sin t\,d\alpha_n(t) \right\|$$

$$= \| (1-\rho_{1,n}) \cos x \| = (1-\rho_{1,n}) = 2\sin^2 (2n)^{-1}$$

Since $(\sin^2(2n)^{-1})\sim(n^{-2})$ condition (3.1.6) is satisfied.

Example 2. We modify the operators A_n of Example 1 slightly. For $n \in N$ let

$$d\bar{\alpha}_n = \frac{\pi}{2}(1 - \frac{1}{n})(d\rho_{-n^{-1}} + d\rho_{n^{-1}}) + \frac{\pi}{2n}(d\rho_{-\pi} + d\rho_{\pi})$$

and $\bar{A}_n(f) = f * d\bar{\alpha}_n$. Thus, we have altered $d\alpha_n$ by the addition of masses $(2n)^{-1}\pi$ at the end points $-\pi$ and π . For the Fourier-Stieljes coefficients $\bar{\rho}_{k,n}$ of $d\bar{\alpha}_n$, we now have

$$1-\bar{\rho}_{k,n} = \int_{-\pi}^{\pi} (1-\cos kt)\, d\bar{\alpha}_n(t)$$

$$= 2(1 - \frac{1}{n})\sin^2 \frac{k}{2n} + \frac{2}{n} \sin^2 \frac{k\pi}{2} \tag{3.2.1}$$

In particular,

$$1-\bar{\rho}_{1,n} = 2\, n^{-1} + o(n^{-1})$$

and

$$1-\bar{\rho}_{2,n} = 2n^{-2} + o(n^{-2})$$

Therefore, the sequence $(\bar{A}_n)$ is not saturated with order $(1-\bar{\rho}_{1,n})\sim(n^{-1})$ since the function $\cos 2t$ is not constant but nevertheless approximated with order $O(n^{-2}) = o(1-\bar{\rho}_{1,n})$. This estimate is obtained the same way we have estimated for $\cos t$ in Example 1 .

Instead the sequence $(\bar{A}_n)$ is saturated with order $(1-\bar{\rho}_{2,n}) \sim (n^{-2})$. To see this, first suppose $f \in C^*$ and for some subsequence (n_j)

$$\|f-\bar{A}_{n_j}(f)\| = o(n_j^{-2})$$

then for $k = \pm 1, \pm 2, \ldots$

$$\hat{f}(k)(1-\bar{\rho}_{k,n_j}) = o(n_j^{-2})$$

This implies $\hat{f}(k) = 0, k = \pm 1, \pm 2, \ldots,$ by virtue of (3.2.1). Therefore, f is constant.

The function $f_o(t) = \cos 2t$, as we have mentioned above, satisfies

$$\| f_o-\bar{A}_n(f_o)\| = O(n^{-2}) .$$

This example points out that in general it is necessary to consider Fourier-Stieljes coefficients of order higher than 1 to determine the saturation properties of a sequence of convolution operators.

3.3. A Necessary and Sufficient Condition for Positive Convolution to be Saturated. It should be clear from the above examples that if (L_n) is to be saturated with order $(1-\rho_{m,n})$ then for each $k \in N$ we must have

$$\varliminf_{n\to\infty} \frac{1-\rho_{k,n}}{1-\rho_{m,n}} > 0$$

since otherwise the corresponding function $\cos kx$ would contradict (3.1.5).

We now show that this condition actually allows us to characterize when convolution operators are saturated.

Theorem 3.1. *Let (L_n) be a sequence of operators of the form* (3.1.1). *A necessary and sufficient condition that (L_n) be saturated is that for some positive integer m*

$$\lim_{n\to\infty} \frac{1-\rho_{k,n}}{1-\rho_{m,n}} = \psi_k > 0 , \quad k \in N \tag{3.3.1}$$

In the case that (3.3.1) *holds $(1-\rho_{m,n})$ is a saturation order.*

Proof. First, suppose that (3.3.1) holds for some $m > 0$. Let $f \in C^*$ and suppose for some subsequence (n_j)

$$\| f-L_{n_j}(f) \| = o(1-\rho_{m,n_j}) \qquad (j\to\infty)$$

then for $k = \pm 1, \pm 2, \ldots$

$$\hat{f}(k)(1-2\check{\mu}_{n_j}(k)) = o(1-\rho_{m,n_j}) \qquad (j\to\infty)$$

In other words,

$$\hat{f}(k)(1-\rho_{|k|,n_j}) = o(1-\rho_{m,n_j}) \qquad (j\to\infty)$$

This together with (3.3.1) implies that $\hat{f}(k) = 0, k = \pm 1, \pm 2, \ldots$. This shows that f is constant and condition (3.1.5) is satisfied.

For the function $f_0(t) = \cos mt$, we have for $n \in N$

$$\| f_o - L_n(f_o) \| = \| \frac{1}{\pi} \int_{-\pi}^{\pi} (\cos mx - \cos m(t+x)) d\mu_n(t) \|$$

$$= \| \frac{1}{\pi} \int_{-\pi}^{\pi} \cos mx (1-\cos mt) d\mu_n(t) \| = \| (1-\rho_{m,n}) \cos mx \| = 1-\rho_{m,n}$$

Therefore, $\cos mt$ is a function which satisfies condition (3.1.6). This establishes the sufficiency of (3.3.1).

To show that (3.3.1) is necessary, we suppose that (L_n) is saturated with order (ϕ_n). For each $k \in N$ and subsequence (n_j) we must have

$$1-\rho_{k,n_j} \neq o(\phi_{n_j})$$

since otherwise the non-constant function $\cos kt$ would contradict (3.1.5). (Here, we would estimate exactly the same way as we estimated for $\cos mt$ above). Thus, for $k \in N$

$$\eta_k = \varliminf_{n \to \infty} \frac{1-\rho_{k,n}}{\phi_n} > 0 \qquad (3.3.2)$$

We now show that for some $m \in N$, $1-\rho_{m,n} = O(\phi_n)$ $(n \to \infty)$. Let $f_o \in C^*$ be a non-constant function such that

$$\| f_o - L_n(f_o) \| = O(\phi_n) . \qquad (3.3.3)$$

Since f_o is not constant, for some $j \neq 0, \hat{f}_o(j) \neq 0$. But then from (3.3.3)

$$\hat{f}_o(j)(1-\rho_{|j|,n}) = O(\phi_n)$$

and so

$$\frac{1-\rho_{|j|,n}}{\phi_n} = O(1)(n\to\infty) \tag{3.3.4}$$

Therefore, we can take $m = |j|$, so that using (3.3.4) together with (3.3.2), we have $1-\rho_{m,n} = O(\phi_n)$,$m = |j|$. Finally, using this with (3.3.2) we have for $k \in N$

$$\varliminf_{n\to\infty} \frac{1-\rho_{k,n}}{1-\rho_{m,n}} > 0$$

which shows that (3.3.1) is necessary. Also $(\phi_n) \sim (1-\rho_{m,n})$ and thus $(1-\rho_{m,n})$ is a saturation order.

3.4. An Example of a Non-saturated Sequence of Positive Convolution Operators

Let's give an example of a sequence of positive convolution operators which is not saturated. The idea is to construct measures $d\mu_n$ such that for each $k \in N$

$$1-\rho_{2k,n} = o(1-\rho_{k,n})(n\to\infty) .$$

This would prevent condition (3.3.1) from holding. For $k = 0,1,2, \ldots,$ let $x_k = 2^{-k}\pi$ and

$$d\mu_n = a_n\left(d\rho_o + \sum_{k=o}^{\infty} (k+1)^{-2}\, n^{-2+\frac{1}{k+1}} [d\rho_{x_k} + d\rho_{-x_k}]\right)$$

where a_n is chosen so that $\frac{1}{\pi}\int_{-\pi}^{\pi} d\mu_n(t) = 1$. The sequence (L_n) given by $L_n(f) = f*d\mu_n$, $n \in N$, is not saturated. For suppose (3.3.1) was satisfied for some m . We write $m = 2^{\alpha}m_o$, where α is a non-negative integer , $m_o \in N$ is not divisible by 2 . Then ,

$$1-\rho_{m,n} = a_n \frac{4}{\pi} \sum_{k=0}^{\infty} (k+1)^{-2} n^{-2+\frac{1}{k+1}} \sin^2(2^{\alpha-k-1} m_0 \pi)$$

$$= \frac{4a_n}{\pi} \sum_{k=\alpha}^{\infty} (k+1)^{-2} n^{-2+\frac{1}{k+1}} \sin^2 (2^{\alpha-k-1} m_0 \pi)$$

$$= \frac{4a_n}{\pi} (\alpha+1)^{-2} n^{-2+\frac{1}{\alpha+1}} \sin^2 \frac{m_0 \pi}{2} + o(a_n n^{-2+\frac{1}{\alpha+1}})$$

A similar estimate gives

$$1-\rho_{2m,n} = \frac{4a_n}{\pi} (\alpha+2)^{-2} n^{-2+\frac{1}{\alpha+2}} \sin^2 \frac{m_0 \pi}{2} + o(a_n n^{-2+\frac{1}{\alpha+2}})$$

Therefore,

$$1-\rho_{2m,n} = o(1-\rho_{m,n})$$

which contradicts (3.3.1).

3.5. Direct and Inverse Theorems for Characterizing the Saturation Class

We have just given necessary and sufficient conditions(in terms of Fourier-Stieljes coefficients) for determining when (L_n) is saturated. The more difficult part of the saturation problem is to determine the saturation class. In this and the next several sections, we will discuss the problem of characterizing the saturation class. We at first restrict our attention to the case when the saturation order is determined by the first Fourier-Stieljes coefficients. In section 3.12, we will describe the necessary modifications to handle the general case when $(1-\rho_{m,n})$ determines the saturation order.

In 1941, G. Alexits[1] in a now classical theorem gave a characterization of the saturation class for Fejèr operators (the "o" saturation theorem for

Fejer operators goes back to A. Zygmund). This paper of Alexits was the beginning of the study of saturation of convolution operators. In 1949, J. Favard$_{1,2}$ gave a general formulation of the phenomenon of saturation for convolution operators. Later, M. Zamanski$_1$ initiated a systematic study of saturation for convolution with trigonometric polynomials. In particular, Zamanski was able to recover the Alexits theorem.

The Fourier transform method for characterizing the saturation class was initiated in the papers of P.L. Butzer$_{1,2,3}$ (for singular integrals on the line), and G. Sunouchi and C. Watari$_1$ (see also F. Harsiladge$_1$). The latter is a good illustration of the transform technique. In this theorem, it is not necessary to assume the operators are positive. However, we still require that the measures are even and normalized.

Theorem 3.2. *Let (L_n) be a sequence of convolution operators (not necessarily positive) and suppose that for each* $k \in N$

$$\lim_{n\to\infty} \frac{1-\rho_{k,n}}{1-\rho_{1,n}} = \Psi_k \neq 0 \tag{3.5.1}$$

Then (L_n) is saturated with order $(1-\rho_{1,n})$.

If $f \in S(L_n)$ *then*

$$\sum_{k=-\infty}^{\infty} \Psi_{|k|} \hat{f}(k)e^{ikx} = \sum_{1}^{\infty} \Psi_k A_k(f,x) \in L_\infty \tag{3.5.2}$$

where we define $\Psi_0 = 0$.

Proof. We first establish that (L_n) is saturated with order $(1-\rho_{1,n})$. If for some $f \in C^*$,

$$\lim_{n\to\infty} \|(1-\rho_{1,n})^{-1}(f-L_n(f))\| = 0$$

then for some subsequence (n_j)

$$\lim_{j\to\infty} \|(1-\rho_{1,n_j})^{-1}(f-L_{n_j}(f))\| = 0$$

Thus, for each $k = \pm 1, \pm 2, \ldots$

$$\hat{f}(k)\Psi_{|k|} = \lim_{j\to\infty} \hat{f}(k) \frac{1-\rho_{k,n_j}}{1-\rho_{1,n_j}} = 0$$

so that $\hat{f}(k) = 0$, $k = \pm 1, \pm 2, \ldots$. This gives that f is constant. The function $f_0(t) = \cos t$ is a non-constant function approximated with order $(1-\rho_{1,n})$.

To prove (3.5.2) we suppose $f \in S(L_n)$. Then

$$\| f-L_n(f)\| = O(1-\rho_{1,n}) .$$

Therefore, the functions

$$g_n(x) = (1-\rho_{1,n})^{-1}(f(x) - L_n(f,x))$$

all lie in some fixed bounded sphere in $L_\infty[-\pi,\pi]$. Since, L_∞ is the conjugate space of L_1 , any closed bounded sphere in L_∞ is weak * compact. Thus, there is a function $g \in L_\infty$ and a subsequence (n_j) such that g_{n_j} converges weak * to g . The function e^{ikx} is in L_1 , thus

$$\hat{g}(k) = \lim_{j\to\infty} \frac{1}{2\pi} \int_{-\pi}^{\pi} g_{n_j}(x)\, e^{-ikx}$$

$$= \lim_{j\to\infty} \hat{f}(k)(1-\rho_{1,n_j})^{-1}(1-\rho_{|k|,n_j}) = \Psi_{|k|}\hat{f}(k) .$$

This gives (3.5.2) .

A theorem which gives necessary conditions for a function to be in the saturation class is commonly called an inverse theorem. When sufficient conditions are given, it is called a direct theorem. At this point, it would be very nice if we could prove a direct theorem under the hypothesis of Theorem 3.2. Namely, that if $\sum \Psi_k A_k(f,x) \in L_\infty$ then $f \in S(L_n)$. This would then completely characterize the saturation class. Unfortunately, such a general direct theorem does not hold, as we shall see in Section 3.8. However, if we assume a certain multiplier condition on the Fourier-Stieljes coefficients we can prove a direct theorem which is a companion to Theorem 3.2. Direct theorems using multiplier conditionswere first given by H. Buchwalter[1] and in the case we are interested in, by G. Sunouchi[1]

Theorem 3.3. *Let* (L_n) *be a sequence of convolution operators for which* (3.5.1) *holds. Suppose further, that*

$$(\Psi_k^{-1}(1-\rho_{1,n})^{-1}(1-\rho_{k,n}))_{k=1}^{\infty} \tag{3.5.3}$$

are multipliers in (C^*,C^*) *with norms uniformly bounded in* n . *If* $f \in C^*$ *with*

$$\sum_{k=1}^{\infty} \Psi_k A_k(f,x) \in L_\infty$$

then $f \in S(L_n)$.

Proof. Let $d\gamma_n$ be the even Borel measure on $[-\pi,\pi]$ with mean value 0 which has Fourier-Stieljes coefficients

$$2\check{\gamma}_n(k) = \Psi_k^{-1}\,(1-\rho_{1,n})^{-1}(1-\rho_{k,n})$$

for k, $n \in N$. Then there is a constant M such that

$$\int_{-\pi}^{\pi}|d\gamma_n(t)| \le M\,, \qquad n \in N$$

since we have assumed that the multipliers are uniformly bounded in n (see Section 1.4) . Let $f \in C^*$ with $\hat{f}(0) = 0$ and

$$g \sim \sum_{k=1}^{\infty} \Psi_k\, A_k(f,x) \in L_\infty$$

Then

$$(1-\rho_{1,n})^{-1}(f-L_n(f)) = g*d\gamma_n \,.$$

Therefore,

$$\|(1-\rho_{1,n})^{-1}(f-L_n(f))\| \le \|g\|_\infty \int_{-\pi}^{\pi}|d\gamma_n| \le M\|g\|_\infty$$

which shows that $f \in S(L_n)$. When $\hat{f}(0) \neq 0$, then $f - \hat{f}(0)$ is in $S(L_n)$ so $f \in S(L_n)$, since constants are preserved by L_n .

3.6. Saturation of the Fejér Operators and the Typical Means of the Fourier Series. As an example, let's determine the saturation of the Fejér operators, which is due to G. Alexits$_1$ (see also M. Zamanski$_1$)

Theorem 3.4. *The Fejér operators σ_n are saturated with order (n^{-1}) and saturation class*

$$S(\sigma_n) = \{f : \tilde{f} \in \mathrm{Lip}\, 1\} .$$

Proof. We have

$$\sigma_n(f) = f*F_n$$

with $F_n(t) = \frac{1}{2} + \sum_1^n (1 - \frac{k}{n+1})\cos kt$. Therefore,

$$1 - \rho_{k,n} = \begin{cases} \frac{k}{n+1} , & k \leq n+1 \\ 1 , & k \geq n+1 \end{cases}$$

Thus, for $k \in N$

$$\lim_{n\to\infty} n(1-\rho_{k,n}) = k$$

which gives, by virtue of Theorem 3.2, that (σ_n) is saturated with order (n^{-1}) . Also, if $f \in S(L_n)$

then

$$\sum_{k=1}^{\infty} k\ A_k(f,x) \ \in L_\infty \tag{3.6.1}$$

We see from Theorem 1.9 that (3.6.1) is equivalent to $\tilde{f} \in \text{Lip } 1$.

Now to show that each $f \in C^*$ with $\tilde{f} \in \text{Lip } 1$ is in $S(L_n)$ we use Theorem 3.3 . We need to show that the sequences $(\gamma_k(n))_{k=1}^{\infty}$ given by

$$\gamma_k(n) = \begin{cases} 1 \ , & k \le n+1 \\ \dfrac{n+1}{k} \ , & k \ge n+1 \end{cases}$$

are multipliers in (C^*,C^*) with norms uniformly bounded in n . This will follow from the fact that $(\gamma_k(n))$ is quasi-convex uniformly in n (see 1.4.8) . We have

$$\gamma_k(n) - 2\gamma_{k+1}(n) + \gamma_{k+2}(n) = \begin{cases} 0 \ , & k \le n-1 \\ -\dfrac{1}{n+2} \ , & k = n \\ \dfrac{2(n+1)}{k(k+1)(k+2)} \ , & k \ge n+1 \end{cases}$$

Therefore, for each $n \in N$

$$\sum_{k=0}^{\infty} (k+1)\left|\gamma_k(n) - 2\gamma_{k+1}(n) + \gamma_{k+2}(n)\right| = \frac{n+1}{n+2} + \sum_{k=n+1}^{\infty} \frac{2(n+1)}{k(k+2)}$$

$$\leq 1 + \sum_{j=1}^{\infty} \Big(\sum_{nj+1}^{n(j+1)} \frac{2(n+1)}{k^2} \Big) \leq 1 + \sum_{j=1}^{\infty} \frac{2n(n+1)}{(nj)^2}$$

$$\leq 1 + 2(1 + \tfrac{1}{n}) \sum_{j=1}^{\infty} j^{-2} \leq C$$

as desired.

More generally we can consider the typical means of the Fourier series of order α, with $0 < \alpha$. Let

$$R_n^{\alpha}(f,x) = \frac{1}{2} + \sum_{1}^{n} \Big(1 - \big(\frac{k}{n+1}\big)^{\alpha}\Big) A_k(f,x) = f * \underline{X}_{\alpha}$$

Then for $0 < \alpha < 2$, (R_n^{α}) is a sequence of positive operators. The saturation of (R_n^{α}) was given by S. Aljancic[1] (see also Sunouchi and Watari[1])

<u>Theorem 3.5. If $\alpha > 0$ then the sequence (R_n^{α}) is saturated with order $(n^{-\alpha})$ and saturation class</u>

$$S(R_n^{\alpha}) = \{f : \sum_{1}^{\infty} k^{\alpha} A_k(f,x) \in L_{\infty}\}$$

<u>Proof.</u> It is clear that

$$1 - \rho_{k,n} = \begin{cases} \left(\frac{k}{n+1}\right)^{\alpha}, & \text{if } k \leq n \\ 1, & \text{if } k > n \end{cases}$$

Thus

$$\lim_{n\to\infty} \frac{1-\rho_{k,n}}{1-\rho_{1,n}} = k^{\alpha} .$$

This gives that (R_n^{α}) is saturated with order $(n^{-\alpha})$. In order to characterize the saturation class, we need to show that if

$$\gamma_k(n) = \begin{cases} 1 & , \quad k \leq n \\ (\frac{n+1}{k})^{\alpha} & , \quad k > n \end{cases}$$

then $\gamma_k(n)$ is quasi convex uniformly in n. To see this, we first find that

$$\gamma_k(n)-2\gamma_{k+1}(n)+\gamma_{k+2}(n) = \begin{cases} 0 & , \quad k \leq n-1 \\ -1 + (\frac{n+1}{n+2})^{\alpha} & , \quad k = n \\ (n+1)^{\alpha}[k^{-\alpha}-2(k+1)^{-\alpha}+(k+2)^{-\alpha}] & , \quad k \geq n+1 \end{cases} .$$

Now, if $h(x) = x^{\alpha}$, there are points ξ_k, ξ_k', $\eta_k \in [(k+2)^{-1},k^{-1}]$ such that

$$|k^{-\alpha} - 2(k+1)^{-\alpha} + (k+2)^{-\alpha}| \leq |\frac{h'(\xi_k)}{k(k+1)} - \frac{h'(\xi_k')}{(k+1)(k+2)}|$$

$$\leq (k+1)^{-1}(k+2)^{-1} \; |\xi_k-\xi_k'| \; |h''(\eta_k)| + 2k^{-1}(k+1)^{-1}(k+2)^{-1} \; |h'(\xi_k)|$$

$$\leq 2k^{-4} \; \alpha(\alpha-1)(k+2)^{2-\alpha} + 2k^{-3} \; \alpha(k+2)^{1-\alpha} \leq Ck^{-2-\alpha}$$

where C is a constant independent of k. Finally,

$$\sum_{k=0}^{\infty} (k+1) \; |\gamma_k(n) - 2\gamma_{k+1}(n) + \gamma_{k+2}(n)|$$

$$\leq n[1 - (\frac{n+1}{n+2})^{\alpha}] + C(n+1)^{\alpha} \sum_{k=n+1}^{\infty} k^{-1-\alpha} \leq M$$

where to obtain a bound for the first term we use the mean value theorem on the function x^{α}. To estimate the second term we sum the series in the same way as we did above for the Fejer operators.

3.7. <u>Tureckiĭ's Theorem.</u> As was observed in Section 3.5, the inverse theorem 3.2 and the direct theorem 3.3 give a characterization of the saturation class only when the multiplier condition (3.5.3) holds. G. Sunouchi[3] has constructed examples of convolution operators where (3.5.1) holds, but the multiplier condition (3.5.3) is not valid. Sunouchi has shown that if $\alpha > 0$ then there is a sequence (h_n) of even functions in $L_1[-\pi,\pi]$, with uniformly bounded L_1 norms such that

$$\lim_{n\to\infty} n^{\alpha}(1-2\hat{h}_n(k)) = k^{\alpha} \qquad k \in N$$

and yet the norms of the multipliers

$$\left(\frac{n^{\alpha}(1-2\hat{h}_n(k))}{k^{\alpha}}\right)_{k=1}^{\infty}$$

tend to infinity with n. From this, we can deduce (see Section 3.8) that there is a function $f_0 \in C^*$ such that

$$\sum_{k=1}^{\infty} k^{\alpha} A_k(f_0,x) \in L_{\infty} \tag{3.7.1}$$

and

$$\| f_o - f_o * h_n \| \neq O(n^{-\alpha})$$

Thus, the saturation class for the operators $L_n(f) = f*h_n$ is not characterized by (3.7.1) .

When the measures are positive, the situation changes. For one thing, the operators in Sunouchi's example are not positive. Secondly, the case $\alpha > 2$ does not occur since

$$1-\rho_{k,n} = \frac{2}{\pi} \int_{-\pi}^{\pi} \sin^2 \frac{kt}{2} \, d\mu_n(t) \leq$$

$$\leq \frac{2k^2}{\pi} \int_{-\pi}^{\pi} \sin^2 \frac{t}{2} \, d\mu_n(t) = k^2(1-\rho_{1,n})$$

That is to say, if the measures $d\mu_n$ are postive and satisfy (3.5.1) then $\Psi_k = O(k^2)$.

Moreover, in the case that $\Psi_k = k^2$, then A.H. Tureckiĭ$_1$ has shown that (3.7.1) does characterize the saturation class.

<u>Theorem 3.6.</u> <u>Let</u> (L_n) <u>be a sequence of positive convolution operators of the form</u> (3.1.1). <u>If for each</u> $k \in N$

$$\lim_{n\to\infty} \frac{1-\rho_{k,n}}{1-\rho_{1,n}} = k^2 \tag{3.7.2}$$

then (L_n) is saturated with order $(1-\rho_{1,n})$ and saturation class

$$S(L_n) = \{f : \sum_1^{\infty} k^2 A_k(f,x) \in L_\infty\} = \{f : f' \in \mathrm{Lip}\ 1\}$$

Proof. From Theorem 3.2, it follows that (L_n) is saturated with order $(1-\rho_{1,n})$ and for each $f \in S(L_n)$ we have

$$\sum_{k=1}^{\infty} k^2 A_k(f,x) \in L_\infty$$

Thus for each $f \in S(L_n)$, $f' \in \text{Lip } 1$ (see Theorem 1.9) .

If $f' \in \text{Lip } 1$ then the estimate (2.4.2) shows that

$$\| f - L_n(f) \| = O(1-\rho_{1,n})$$

and thus $f \in S(L_n)$.

3.8. Examples when the direct theorem does not hold. The importance of Tureckii's theorem vis a vis Theorems 3.2 and 3.3, is that it is not necessary to assume the multiplier condition (3.5.3). We will examine Tureckiĭ's theorem more closely in the next few sections, but first, we wish to clarify the nature of the multiplier condition (3.5.3) by giving a general procedure (see De Vore$_5$) for constructing examples of operators $L_n(f) = f*d\mu_n$ where (3.5.1) holds and yet (3.5.2) does not characterize the saturation class. Because of Tureckii's theorem we must avoid the case $\Psi_k = k^2$. We do this by assuming

$$\Psi_k = o(k^2) \tag{3.8.1}$$

It is not necessary to assume that the $d\mu_n$ are positive, but only

$$1-\rho_{k,n} > 0 \qquad k , n \in N \tag{3.8.2}$$

which is clearly satisfied when $d\mu_n$ is positive. We also need to assume that

$$(\Psi_k^{-1}) \text{ is a multiplier from } C^* \text{ to } C^* \tag{3.8.3}$$

These conditions are satisfied for the sequence $(\Psi_k) = (k^\alpha)$ whenever $0 < \alpha < 2$. The latter since $(k^{-\alpha})$ is quasi convex and tends to 0.

Theorem 3.7. Let (L_n) be a sequence of convolution operators, $L_n(f) = f*d\mu_n$ which satisfy (3.5.1) with (Ψ_k) satisfying (3.8.1), (3.8.2), and (3.8.3). Then there is a sequence of positive numbers (ξ_n) converging to 0 such that the modified measures

$$d\bar{\mu}_n(t) = \frac{1}{2}[d\mu_n(t-\xi_n) + d\mu_n(t+\xi_n)] \tag{3.8.4}$$

satisfy for $k \in N$,

$$\lim_{n\to\infty} \frac{1-\bar{\rho}_{k,n}}{1-\rho_{1,n}} = \Psi_k \tag{3.8.5}$$

where $\bar{\rho}_{k,n}$ are the coefficients of $d\bar{\mu}_n$. There is a function $f_o \in C^*$ with

$$\sum_1^\infty \Psi_k A_k(f_o,x) \in L_\infty$$

and

$$\| f_o - f_o * d\bar{\mu}_n \| \neq O(1-\rho_{1,n})$$

<u>Proof.</u> Let $a_n = \max\{m^{-2}\Psi_m : m \geq (1-\rho_{1,n})^{-\frac{1}{2}}\}$. Then $a_n \to 0$ because of (3.8.1). Define $\xi_n = a_n^{\frac{1}{4}}(1-\rho_{1,n})^{\frac{1}{2}}$ and let $d\bar{\mu}_n$ be defined by (3.8.4). Then an easy computation shows

$$\bar{\rho}_{k,n} = \frac{1}{\pi}\int_{-\pi}^{\pi} \cos kt \; d\bar{\mu}_n(t) = \cos k\xi_n \, \rho_{k,n} \ .$$

Thus,

$$\frac{1-\bar{\rho}_{k,n}}{1-\rho_{1,n}} = \frac{1-\cos k\xi_n \rho_{k,n}}{1-\rho_{1,n}} = \frac{2\sin^2 \frac{k\xi_n}{2}}{1-\rho_{1,n}} + \cos k\xi_n \frac{1-\rho_{k,n}}{1-\rho_{1,n}}$$

Since

$$\sin^2 \frac{k\xi_n}{2} = O(\xi_n^2) = o(1-\rho_{1,n}) \ ,$$

we have

$$\lim_{n\to\infty} \frac{1-\bar{\rho}_{k,n}}{1-\rho_{1,n}} = \lim_{n\to\infty} \cos k\xi_n \frac{1-\rho_{k,n}}{1-\rho_{1,n}} = \Psi_k$$

which shows (3.8.5) .

Now we want to establish the existence of the function f_o . Let $m_n = [\xi_n^{-1}] + 1$. Then $\xi_n^{-1} \leq m_n \leq \xi_n^{-1} + 1$. Therefore,

$$\frac{1-\bar{\rho}_{m_n,n}}{\Psi_{m_n}(1-\rho_{1,n})} = \frac{2\sin^2 \frac{m_n\xi_n}{2}}{\Psi_{m_n}(1-\rho_{1,n})} + \frac{\cos m_n\xi_n(1-\rho_{m_n,n})}{\Psi_{m_n}(1-\rho_{1,n})}$$

Now $(1-\rho_{1,n})^{-1}(1-\rho_{m_n,n})$ is positive from (3.8.2) and $\cos m_n\xi_n \to \cos 1 > 0$ so that for n sufficiently large

$$\frac{1-\bar{\rho}_{m_n,n}}{\Psi_{m_n}(1-\rho_{1,n})} \geq \frac{\sin^2 \frac{m_n\xi_n}{2}}{\Psi_{m_n}(1-\rho_{1,n})} \geq \frac{\sin^2 \frac{1}{2}}{\Psi_{m_n}(1-\rho_{1,n})} .$$

Since

$$m_n \geq \xi_n^{-1} \geq (1-\rho_{1,n})^{-\frac{1}{2}} ,$$

we have

$$\Psi_{m_n}(1-\rho_{1,n}) \leq a_n m_n^2 (1-\rho_{1,n}) \leq a_n(1+\xi_n^{-1})^2 (1-\rho_{1,n})$$

$$\leq a_n(a_n^{-\frac{1}{4}}((1-\rho_{1,n})^{-\frac{1}{2}} + 1)^2 (1-\rho_{1,n}) \to 0 .$$

This shows that

$$\frac{1-\bar{\rho}_{m_n,n}}{\Psi_{m_n}(1-\rho_{1,n})} \to \infty . \qquad (3.8.6)$$

Let $d\gamma_n$ be the even Borel measure with mean value 0 whose complex Fourier coefficients are

$$\check{\gamma}_n(k) = \frac{1}{2}\,\frac{1-\bar{\rho}_{k,n}}{\Psi_k(1-\rho_{1,n})} \qquad k = \pm 1 , \pm 2 , \dots .$$

By virtue of (3.8.6), we must have

$$\int_{-\pi}^{\pi} |d\gamma_n(t)| \to \infty .$$

Therefore, using the uniform boundeness principle, there must be a function $g \in C^*$ such that $\| g * d\gamma_n \| \to \infty$.

Let f_o be the function in C^* with Fourier coefficient

$$\hat{f}_o(k) = \Psi_k^{-1} \hat{g}(k) .$$

The existence of such a function is guaranteed by (3.8.3) . Checking Fourier coefficients we see that

$$\frac{f_o - f_o * d\bar{\mu}_n}{1-\rho_{1,n}} = g * d\gamma_n$$

Thus

$$\|f_o - f_o * d\bar{\mu}_n\| \neq O(1-\rho_{1,n})$$

However

$$\sum_1^{\infty} \Psi_k A_k(f_o, x) = \sum_1^{\infty} A_k(g,x) \in L_{\infty}$$

which proves the theorem.

One interesting example occurs when we let $d\mu_n(t) = F_n(t)dt$ where F_n is the Fejér kernel. Then the Fejer operators $\sigma_n(f) = f * F_n$ are saturated with order (n^{-1}) as we have seen in Section 3.6. A simple check shows that we can take $\xi_n = n^{-5/8}$. Thus the operators

$$L_n(f) = f*\Lambda_n$$

where

$$\Lambda_n(t) = \frac{1}{2}[F_n(t-n^{-5/8}) + F_n(t+n^{-5/8})]$$

are saturated with order (n^{-1}) and the Fourier coefficients of (Λ_n) satisfy

$$\lim_{n\to\infty} n(1-2\hat{\Lambda}_n(k)) = k .$$

Yet the saturation class $S(L_n) \neq \{f : \sum_1^\infty k\ A_k(f,x) \in L_\infty\}$.

More generally, we can consider the typical means of the Fourier series, for $0 < \lambda < 2$, introduced in Section 3.6.

$$R_n^\lambda(f,x) = \frac{1}{2} + \sum_1^n (1-(\frac{k}{n+1})^\lambda)\ A_k(f,x) = f*X_n^\lambda$$

which are saturated with order $(n^{-\lambda})$. The saturation class for (R_n^λ) is

$$S(R_n^\lambda) = \{f : \sum_1^\infty k^\lambda\ A_k(f,x) \in L_\infty\} .$$

The sequence $(k^{-\lambda})$ is a multiplier from C^* to C^* . In this case, we can take

$$\xi_n^{(\lambda)} = n^{\frac{\lambda^2}{8} - \frac{3\lambda}{4}}$$

Then, the modified operators

$$L_n^\lambda(f) = f * \Lambda_n^\lambda$$

where

$$\Lambda_n^\lambda(t) = \frac{1}{2}[X_n^\lambda(t-\xi_n^{(\lambda)}) + X_n^\lambda(t+\xi_n^{(\lambda)})]$$

satisfy

$$\lim_{n\to\infty} n^\lambda(1-2\hat{\Lambda}_n^\lambda(k)) = k^\lambda$$

and yet

$$S(L_n^\lambda) \neq \{f : \sum_1^\infty k^\lambda A_k(f,x) \in L_\infty\}$$

3.9. Equivalent Formulations of Tureckiĭ's Theorem. There are several equivalent formulations of Tureckiĭ's Theorem which center about replacing (3.7.2) by an equivalent condition. We summarize the important ones in the following theorem.

Theorem 3.8. *Let (L_n) be a sequence of positive convolution operators of the form* (3.1.1) . *Then, the following are equivalent*

3.9.1. *For each* $k \in N$

$$\lim_{n\to\infty} \frac{1-\rho_{k,n}}{1-\rho_{1,n}} = k^2$$

3.9.2.

$$\lim_{n\to\infty} \frac{1-\rho_{2,n}}{1-\rho_{1,n}} = 4$$

3.9.3.

$$\int_{-\pi}^{\pi} \sin^4 \frac{t}{2}\, d\mu_n(t) = o(1-\rho_{1,n})$$

3.9.4. For each $\delta > 0$

$$\int_{\delta}^{\pi} d\mu_n(t) = o(1-\rho_{1,n})$$

The equivalence of (3.9.1) and (3.9.2) was first noted by Korovkin_3. Asymptotic conditions like (3.9.3) were first studied by E. Komleva_1.
The other equivalences were given by E. Görlich and E. $\text{Stark}_{1,2}$. I.P. Natanson_1 has used (3.9.3) in place of (3.9.1) in his version of Tureckiĭ's Theorem.

Proof of Theorem 3.8. It is clear that (3.9.1) implies (3.9.2). Let us show that (3.9.2) implies (3.9.3). Rewriting (3.9.2), we have

$$\int_{-\pi}^{\pi} \sin^2 t\, d\mu_n(t) - 4 \int_{-\pi}^{\pi} \sin^2 \frac{t}{2}\, d\mu_n(t) = o(1-\rho_{1\ n})$$

Owing to the identity

$$\sin^2 t - 4 \sin^2 \frac{t}{2} = 4 \sin^2 \frac{t}{2} \cos^2 \frac{t}{2} - 4 \sin^2 \frac{t}{2} = -4 \sin^4 \frac{t}{2}$$

we have

$$4 \int_{-\pi}^{\pi} \sin^4 \frac{t}{2}\, d\mu_n(t) = o(1-\rho_{1,n})$$

which is (3.9.3). Since

$$\int_{\delta}^{\pi} d\mu_n(t) \leq \frac{1}{\sin^4 \frac{\delta}{2}} \int_{-\pi}^{\pi} \sin^4 \frac{t}{2}\, d\mu_n(t)$$

we see that (3.9.3) implies (3.9.4).

Finally, we wish to show that (3.9.4) implies (3.9.1). Fix k, then if $\epsilon > o$, choose $\delta > o$ such that

$$\left|\sin^2 \frac{kt}{2} - k^2 \sin^2 \frac{t}{2}\right| \leq \frac{\epsilon}{2} \sin^2 \frac{t}{2} \quad \text{for} \quad |t| \leq \delta$$

Then,

$$\frac{\pi}{2}\left|(1-\rho_{k,n}) - k^2(1-\rho_{1,n})\right| \leq \int_{-\delta}^{\delta} \left|\sin^2 \frac{kt}{2} - k^2 \sin^2 \frac{t}{2}\right| d\mu_n(t)$$

$$+ 2\int_{\delta}^{\pi} \left|\sin^2 \frac{kt}{2} - k^2 \sin^2 \frac{t}{2}\right| d\mu_n(t)$$

$$\leq \frac{\epsilon}{2} \int_{-\delta}^{\delta} \sin^2 \frac{t}{2}\, d\mu_n(t) + 2(1+k^2) \int_{\delta}^{\pi} d\mu_n(t)$$

$$\leq \frac{\epsilon\pi}{4}(1-\rho_{1,n}) + 2(1+k^2) \int_{\delta}^{\pi} d\mu_n(t) .$$

By virtue of (3.9.4), for n sufficiently large say $n > N$

$$2(1+k^2) \int_{\delta}^{\pi} d\mu_n(t) \leq \frac{\epsilon\pi}{4}(1-\rho_{1\,n}) .$$

Thus, for $n \geq N$

$$|(1-\rho_{k,n}) - k^2(1-\rho_{1,n})| \leq \epsilon(1-\rho_{1,n}) \tag{3.9.5}$$

3.10. Applications of Tureckiǐ's Theorem. First consider the operators A_n introduced in Section 3.2. We saw there that the Fourier-Stieljes coefficients satisfied

$$\lim_{n\to\infty} \frac{1-\rho_{k,n}}{1-\rho_{1,n}} = k^2 .$$

Thus from Tureckii's Theorem we deduce that (A_n) is saturated with order $(1-\rho_{1,n}) \sim (n^{-2})$ and saturation class $S(A_n) = \{f: f' \in \text{Lip } 1\}$ (see pg. 69).

As a second example, we want to determine the saturation class for the Jackson operators. Actually, we will consider a generalization of the operators given by Y. Matsuoka[1] . For $n,p,q \in N$ let

$$K_{n,p,q}(t) = C_{n,p,q} \frac{\sin^{2p}\frac{(n+1)t}{2}}{\sin^{2q}\frac{t}{2}} ,$$

where $C_{n,p,q}$ is chosen so that

$$\frac{1}{\pi}\int_{-\pi}^{\pi} K_{n,p,q}(t)dt = 1$$

When $p = q = 1$, we have $K_{n,1,1}(t) = F_n(t)$, the Fejér kernel. When $p = q = 2$, we have $K_n(t)$, the classical Jackson kernel (see Section 2.1).

In order to determine the saturation properties of the operators generated by these kernels, we shall use the following.

<u>Lemma 3.1. For each $p \geq q \geq 1$, we have</u>

$$\int_{-\pi}^{\pi} \frac{\sin^{2p} \frac{(n+1)t}{2}}{\sin^{2q} \frac{t}{2}} \, dt \sim n^{2q-1} (n\to\infty) . \tag{3.10.1.}$$

<u>Proof.</u> First,

$$\int_{-\pi}^{\pi} \frac{\sin^{2p} \frac{(n+1)t}{2}}{\sin^{2q} \frac{t}{2}} \, dt = 2 \int_{0}^{\pi} \frac{\sin^{2p} \frac{(n+1)}{2} t}{\sin^{2q} \frac{t}{2}} \, dt .$$

For $0 \leq t \leq \pi$, $(\frac{t}{\pi})^{2q} \leq \sin^{2q} \frac{t}{2} \leq (\frac{t}{2})^{2q}$ and therefore

$$\int_{-\pi}^{\pi} \frac{\sin^{2p} \frac{(n+1)}{2} t}{\sin^{2q} \frac{t}{2}} \, dt \sim \int_{0}^{\pi} \frac{\sin^{2p} \frac{(n+1)}{2} t}{(\frac{t}{2})^{2q}} \, dt$$

$$= 2(n+1)^{2q-1} \int_{0}^{\frac{(n+1)}{2}\pi} \frac{\sin^{2p} u}{u^{2q}} \, du \sim n^{2q-1}$$

The last estimate follows from the fact that the integrals

$$\int_{0}^{\frac{(n+1)}{2}\pi} \frac{\sin^{2p} u}{u^{2q}} \, du ,$$

converge as $(n\to\infty)$.

<u>Theorem 3.9. For each $p \geq q \geq 2$, the sequence of operators $(J_{n,p,q})_{n=1}^{\infty}$ defined by $J_{n,p,q}(f) = f * K_{n,p,q}$ for $f \in C^*$ are saturated with order (n^{-2})</u>

<u>and saturation class</u> $S(J_{n,p,q}) = \{f : f' \in \mathrm{Lip}\ 1\}$.

<u>Proof.</u> We will show that condition (3.9.3) of Theorem 3.8 is valid for $K_{n,p,q}$ and the theorem will then follow from Theorem 3.4.

From Lemma 3.1, it follows that

$$C_{n,p,q} \sim n^{-2q+1} \tag{3.10.2}$$

and also

$$1-\rho_{1,n} = \frac{2}{\pi} C_{n,p,q} \int_{-\pi}^{\pi} \frac{\sin^{2p} \frac{(n+1)}{2} t}{\sin^{2q-2} \frac{t}{2}} dt \sim n^{-2q+1} n^{2q-3} = n^{-2}$$

If $q \geq 3$ then (3.9.3) follows easily from Lemma 3.1 since

$$\int_{-\pi}^{\pi} \sin^4 \frac{t}{2} K_{n,p,q}(t)dt = C_{n,p,q} \int_{-\pi}^{\pi} \frac{\sin^{2p} \frac{(n+1)}{2} t}{\sin^{2q-4} \frac{t}{2}}$$

$$\sim n^{-2q+1} n^{2q-5} = n^{-4} = o(n^{-2})$$

If $q = 2$, then

$$\int_{-\pi}^{\pi} \sin^4 \frac{t}{2} K_{n,p,q}(t)dt = C_{n,p,q} \int_{-\pi}^{\pi} \sin^{2p} \frac{(n+1)}{2} t \ dt$$

$$= O(C_{n,p,q}) = O(n^{-3}) = o(n^{-2}) \tag{3.10.3}$$

This establishes condition (3.9.3).

In Section 3.1, we promised an example of a sequence of operators which shows the desirability of the $\underline{\lim}$ in 3.1.5, rather than just $\lim$. Consider the sequence (L_n) which is σ_n the Fejer operators when n is odd and J_n the Jackson operators when n is even. If (ϕ_n) is the sequence which is n^{-1} when n is odd and n^{-2} when n is even, then for (L_n) we have $1-\rho_{1,n} \sim \phi_n$. If we use the $\lim$ definition, then (L_n) is saturated with order (n^{-1}) and also with order (ϕ_n). But when we use (ϕ_n) as the saturation order, the saturation class is $\{f : f' \in \text{Lip } 1\}$. When we use (n^{-1}) as the saturation order, the saturation class contains each f for which $f \in \text{Lip } 1$ and $\tilde{f} \in \text{Lip } 1$ and so we see that the saturation classes are not the same.

When we use the definition of saturation with the $\underline{\lim}$ then (L_n) is saturated with order (ϕ_n) but is no longer saturated with order (n^{-1}). The saturation class is then well determined (independent of the saturation order), and is $\{ f : f' \in \text{Lip } 1\}$.

3.11. Generalizations of Tureckiĭ's Theorem. As we have seen, the assumption in Tureckii's theorem (Theorem 3.6) is equivalent to (see Theorem 3.8)

$$\int_{\complement\theta} d\mu_n = o(1-\rho_{1,n}) \tag{3.11.1}$$

for each open set θ containing 0. This condition is, theoretically speaking, very restrictive. For example, if $(d\mu_n)$ is a sequence of measures that satisfy (3.11.1), then if we alter $d\mu_n$ slightly by defining

$$d\bar{\mu}_n(t) = \rho_{1,n} d\mu_n(t) + \frac{1}{2}(1-\rho_{1,n})[d\mu_n(t-\tfrac{\pi}{2}) + d\mu_n(t+\tfrac{\pi}{2})]$$

the new sequence $(d\bar{\mu}_n)$ does not satisfy (3.11.1). However, it is not difficult to show that the operators obtained by convolution with $d\overline{\mu}_n$ still have their saturation class $\{f : f' \in \text{Lip } 1\}$. (This is included in Theorem 3.10).

Of course, the reason (3.11.1) does not hold for $d\overline{\mu}_n$ is that the contribution of $\int d\overline{\mu}_n$ near $\frac{\pi}{2}$ is not $o(1-\rho_{1,n})$. For this reason, we wish to develop another technique for characterizing the saturation class. This technique is particularly applicable to operators whose saturation class is $\{f : f' \in \text{Lip } 1\}$. In terms of Fourier coefficients, we will be able to weaken the assumption (3.7.2) in Tureckiĭ's Theorem to

3.11.2. There is a constant $C > 0$, such that for each $k \in N$, there is an $N(k)$, for which

$$\frac{1-\rho_{k,n}}{1-\rho_{1,n}} \geq Ck^2 \qquad n \geq N(k) .$$

Before we develop the saturation technique we will restate (3.11.2) in terms of the integrals of $d\mu_n$.

Lemma 3.2. If (L_n) is a sequence of positive operators of the form (3.1.1) then (3.11.2) is equivalent to

3.11.3. There is a constant $C' > 0$ such that for each $\epsilon > 0$, there exists an $N(\epsilon)$ so that $n \geq N(\epsilon)$ implies

$$\frac{2}{\pi}\int_{-\epsilon}^{\epsilon} \sin^2 \frac{t}{2}\, d\mu_n(t) \geq C'(1-\rho_{1,n})$$

Proof. We first show that (3.11.3) implies (3.11.2) . Let k be a non-zero integer and choose $0 < \epsilon < \left|\frac{\pi}{k}\right|$. Then

$$\sin^2 \frac{kt}{2} \geq (\frac{2}{\pi})^2 k^2 \sin^2 \frac{t}{2}$$

on $(-\epsilon,\epsilon)$ and so if we let $N(\epsilon)$ be as given in (3.11.3) we have

$$\int_{-\pi}^{\pi} \sin^2 \frac{kt}{2} d\mu_n(t) \geq \int_{-\epsilon}^{\epsilon} \sin^2 \frac{kt}{2} d\mu_n(t) \geq (\frac{2}{\pi})^2 k^2 \int_{-\epsilon}^{\epsilon} \sin^2 \frac{t}{2} d\mu_n(t) \geq$$

$$\geq (\frac{2}{\pi})^2 k^2 C' \int_{-\pi}^{\pi} \sin^2 \frac{t}{2} d\mu_n(t) \quad \text{for} \quad n \geq N(\epsilon) \tag{3.11.4}$$

Therefore, (3.11.2) holds with $N(k) = N(\epsilon)$ and $C = (\frac{2}{\pi})^2 C'$.

Let $\epsilon > 0$ and $S_\epsilon = [-\pi,\pi] \setminus [-\epsilon,\epsilon]$. For $t \in S_\epsilon$,

$$\sin^2 \frac{kt}{2} \leq 1 \leq \sin^{-2} \frac{\epsilon}{2} \sin^2 \frac{t}{2}$$

and so for any $n \in N$

$$\frac{2}{\pi} \int_{S_\epsilon} \sin^2 \frac{kt}{2} d\mu_n(t) \leq \sin^{-2} \frac{\epsilon}{2} (1-\rho_{1,n}) \tag{3.11.5}$$

Choose k so large that $Ck^2 \geq 2 \sin^{-2} \frac{\epsilon}{2}$ and fix k . Because of (3.11.2) and (3.11.5) we have for $n \geq N(k)$

$$k^2 \int_{-\epsilon}^{\epsilon} \sin^2 \frac{t}{2} d\mu_n(t) \geq \int_{-\epsilon}^{\epsilon} \sin^2 \frac{kt}{2} d\mu_n(t) = \int_{-\pi}^{\pi} \sin^2 \frac{kt}{2} d\mu_n(t) - \int_{S_\epsilon} \sin^2 \frac{kt}{2} d\mu_n(t)$$

$$\geq \frac{\pi}{2}(Ck^2 - \sin^{-2} \frac{\epsilon}{2})(1-\rho_{1,n}) \geq \frac{\pi}{4} Ck^2(1-\rho_{1,n}) \tag{3.11.6}$$

Dividing by $\frac{\pi k^2}{2}$ in (3.11.6) establishes (3.11.3) with $C' = \frac{1}{2} C$ and $N(\epsilon) = N(k)$.

We can now give a saturation theorem when $(d\mu_n)$ satisfies either (3.11.2) or (3.11.3) (see DeVore_4)

Theorem 3.10. Let (L_n) be a sequence of positive convolution operators for which either (3.11.2) or (3.11.3) holds. Then (L_n) is saturated with order $(1-\rho_{1,n})$ and saturation class $S(L_n) = \{f : f' \in \text{Lip } 1\}$.

Proof. By virtue of Lemma 3.2, both (3.11.2) and (3.11.3) are satisfied and we will use them interchangeably. We first note that (L_n) is saturated with order $(1-\rho_{1,n})$. This follows from Theorem 3.1 since

$$\varliminf_{n\to\infty} \frac{1-\rho_{k,n}}{1-\rho_{1,n}} \geq C\,k^2 > 0 \quad \text{for} \quad k \in N .$$

We now want to characterize the saturation class $S(L_n)$. As usual, if $f' \in \text{Lip } 1$, by virtue of Theorem 2.4, we have $f \in S(L_n)$.

A function $f \in C^*$ is in $S(L_n)$ if and only if

$$\left\| \int_{-\pi}^{\pi} \frac{f(x+t) + f(x-t) - 2f(x)}{\sin^2 \frac{t}{2}}\, d\Psi_n(t) \right\| = O(1)$$

where

$$d\Psi_n(t) = \frac{1}{\pi}\,(1-\rho_{1,n})^{-1} \sin^2 \frac{t}{2}\, d\mu_n(t) .$$

We need to show that if $f \in S(L_n)$ then $f' \in \text{Lip } 1$. We first wish to show that if f is twice continuously differentiable and

$$\| f-L_n(f) \| \leq M(1-\rho_{1,n}) \tag{3.11.7}$$

then

$$\| f'' \| \leq C(M + \| f \|) \tag{3.11.8}$$

where C is a constant independent of f.

Since each measure $d\Psi_n$ has norm $1/2$ there is a subsequence (n_j) and a measure $d\Psi$ such that $(d\Psi_{n_j})$ converges weak * to $d\Psi$. Using condition (3.11.3) and the weak * convergence we have for each $\epsilon > 0$

$$\int_{-\epsilon}^{\epsilon} d\Psi \geq \lim_{j\to\infty} \int_{-\epsilon}^{\epsilon} d\Psi_{n_j} \geq \frac{C'}{\pi} \tag{3.11.9}$$

Choose ϵ_0 so small that for $T_{\epsilon_0} = [-\epsilon,\epsilon_0] \setminus \{0\}$

$$\int_{T_{\epsilon_0}} d\Psi \leq \frac{C'}{4\pi^2} \tag{3.11.10}$$

Now if f is twice continuously differentiable and satisfies (3.11.7) then

$$\left\| \int_{-\pi}^{\pi} \frac{f(x+t) + f(x-t) - 2f(x)}{\sin^2 \frac{t}{2}} d\Psi(t) \right\| \leq$$

$$\leq \lim_{j\to\infty} \left\| \int_{-\pi}^{\pi} \frac{f(x+t) + f(x-t) - 2f(x)}{\sin^2 \frac{t}{2}} d\Psi_{n_j}(t) \right\| \leq 2M .$$

Thus, letting $S_{\epsilon_0} = [-\pi,\pi] \setminus [-\epsilon_0,\epsilon_0]$

$$\| \int_{-\epsilon_o}^{\epsilon_o} \frac{f(x+t) + f(x-t) - 2f(x)}{\sin^2 \frac{t}{2}} \, d\Psi(t) | \leq$$

$$\leq 2M + \| \int_{S_{\epsilon_o}} \frac{f(x+t) + f(x-t) - 2f(x)}{\sin^2 \frac{t}{2}} \, d\Psi(t) \| \leq$$

$$\leq 2M + \frac{4\| f \|}{\sin^2 \frac{\epsilon_o}{2}} \int_{-\pi}^{\pi} |d\Psi| = 2M + 2\| f \| \sin^{-2} \frac{\epsilon_o}{2} \qquad (3.11.11)$$

Here we know

$$\int_{-\pi}^{\pi} |d\Psi(t)| = \int_{-\pi}^{\pi} d\Psi = \lim_{m\to\infty} \int_{-\pi}^{\pi} d\Psi_m = 1/2 \ .$$

Since $\dfrac{f(x+t) + f(x-t) - 2f(x)}{\sin^2 \frac{t}{2}}$ has the value $4f''(x)$ at $t = 0$, we have from (3.11.9) and (3.11.10)

$$\| \int_{-\epsilon_o}^{\epsilon_o} \frac{f(x+t) + f(x-t) - 2f(x)}{\sin^2 \frac{t}{2}} \, d\Psi(t) \| \geq$$

$$\geq (2-\pi^{-2})C' \| f'' \| - \| \int_{T_{\epsilon_o}} \frac{f(x+t) + f(x-t) - 2f(x)}{\sin^2 \frac{t}{2}} \, d\Psi(t) \| \qquad (3.11.12)$$

But

$$\left| \frac{f(x+t) + f(x-t) - 2f(x)}{\sin^2 \frac{t}{2}} \right| \leq \| f'' \| \frac{t^2}{\sin^2 \frac{t}{2}} \leq \pi^2 \| f'' \| , \quad t > 0$$

and so from (3.11.12) and (3.11.10) it follows that

$$\left\| \int_{-\epsilon_0}^{\epsilon_0} \frac{f(x+t) + f(x-t) - 2f(x)}{\sin^2 \frac{t}{2}} \, d\Psi(t) \right\| \geq$$

$$\geq (2-\pi^{-2}) \quad C' \; \|f''\| - \frac{1}{4} C' \; \|f''\| \geq \quad C' \; \|f''\| \qquad (3.11.13)$$

Using (3.11.13) with (3.11.11) gives

$$\|f''\| \leq \frac{1}{C'}(2M + 2(\sin^{-2} \frac{\epsilon_0}{2})\|f\|)$$

which establishes (3.11.8) .

Finally let f be any function in $S(L_n)$ such that for each $n \in N$

$$\|f - L_n(f)\| \leq M(1-\rho_{1,n})$$

Consider the twice continuously differentiable function $f_m = f{*}K_m$ where K_m is the Jackson kernel of degree $2m$. Then, for f_m we have

$$\|f_m - L_n(f_m)\| = \|f{*}K_m - f{*}K_m{*}d\mu_n\| = \|(f - f{*}d\mu_n){*}K_m\| \leq$$

$$\leq \|f - f{*}d\mu_n\| \; \frac{1}{\pi} \int_{-\pi}^{\pi} K_m(t)\,dt \leq M(1-\rho_{1,n})$$

Hence, from (3.11.7) and the fact that $\|f_m\| \leq \|f\|$, $m \in N$, we have

$$\|f''_m\| \leq C(M + \|f_m\|) \leq C(M + \|f\|)$$

Thus, if $|t| > 0$, $x\epsilon[-\pi,\pi]$, $m \in N$

$$\left| \frac{f_m(x+t) + f_m(x-t) - 2f_m(x)}{t^2} \right| \leq C(M + \|f\|)$$

Taking a limit as $(m\to\infty)$ shows that

$$\left|\frac{f(x+t) + f(x-t) - 2f(x)}{t^2}\right| \leq C(M + \|f\|), \quad x\epsilon[-\pi,\pi] , \; |t| > 0$$

and so $f \in Lip^*2 = \{f : f' \in Lip\ 1\}$ (see Section 1.3) .

3.12. Characterization of the Saturation Class when the Saturation Order is $(1-\rho_{m,n})$. As we have already seen, the saturation of the sequence (L_n) is not always determined by $(1-\rho_{1,n})$. Generally, we must examine higher order Fourier coefficients. In this case, the degree of approximation of f by $(L_n(f))$ is affected by the periodicity as well as the smoothness of f . For example, if the saturation order of (L_n) is $(1-\rho_{2,n})$, than the function cos 2t has a better degree of approximation than the function cos t . This will make the saturation class depend on periodicity as well as smoothness. For this reason, we will need the following estimate for degree of approximation which is in the same spirit as those given in Chapter 2 .

Theorem 3.11. Let (L_n) be a sequence of positive linear operators (not necessarily of convolution type) from C^* into C^* . Let $m \in N$ and suppose (λ_n) is a sequence of positive real numbers such that

$$\| g-L_n(g)\| = O(\lambda_n^2)$$

<u>for the three functions</u> $g(t) = 1$, $\cos mt$, $\sin mt$, <u>Then there is a constant</u> $C > 0$ <u>such that for each</u> $f \in C^*$ <u>of period</u> $\frac{2\pi}{m}$ <u>we have for</u> $n \in N$

$$\| f-L_n(f) \| \leq C(1 + \| f \|)(\lambda_n^2 + \omega(f,\lambda_n)) \qquad (3.12.1)$$

If in addition f is differentiable then

$$\| f-L_n(f) \| \leq C(1 + \| f \| + \| f' \|)(\lambda_n^2 + \lambda_n\ \omega(f',\lambda_n)) \qquad (3.12.2)$$

<u>Proof.</u> This theorem can be proved directly using the same techniques as those given in Chapter 2. However, it is also possible to derive the theorem from Theorem 2.4. Consider the positive linear operators $\bar{L}_n$, $n \in N$ defined by

$$\bar{L}_n(f,x) = L_n(f(mt), \frac{x}{m}) .$$

Then

$$\| 1 - \bar{L}_n(1) \| = \| 1 - L_n(1 , \frac{x}{m}) \| = 0(\lambda_n^2)$$

and

$$\| \cos x - \bar{L}_n(\cos t,x) \| = \| \cos m(\frac{x}{m}) - L_n(\cos mt ,\frac{x}{m}) \| = 0(\lambda_n^2)$$

Similarily,

$$\| \sin x - \bar{L}_n(\sin t,x) \| = 0(\lambda_n^2)$$

Thus, by Theorem 2.4, there is a constant $C > 0$ such that for each $h \in C^*$, $n \in N$

$$\| h - \bar{L}_n(h) \| \leq C(1 + \| h \|)\ (\lambda_n^2 + \omega(h,\lambda_n)) \ .$$

Now, if $f \in C^*$ of period $\frac{2\pi}{m}$, let $h(t) = f(\frac{t}{m})$. Then $h \in C^*$ and $\omega(h,\delta) \leq \omega(f,\frac{\delta}{m}) \leq \omega(f,\delta)$. Therefore, from Theorem 2.4

$$\| h-\bar{L}_n(h) \| \leq C(1 + \| f \|)(\lambda_n^2 + \omega(f,\lambda_n)) \ .$$

Also, for any $x \in [-\pi,\pi]$

$$f(x) - L_n(f,x) = h(mx) - \bar{L}_n(h,mx)$$

Therefore,

$$\| f-L_n(f) \| \leq \| h-\bar{L}_n(h) \| \leq C(1 + \| f \|)(\lambda_n^2 + \omega(f,\lambda_n))$$

In a similar way, we could derive the estimate (3.12.2) for differentiable functions.

If we apply Theorem 3.11 to convolution operators we have

<u>Corollary 3.1</u> <u>Let (L_n) be a sequence of positive convolution operators</u> <u>If $m \in N$, then there is a constant $C_m > 0$ such that for each $f \in C^*$ of period $\frac{2\pi}{m}$</u>

$$\| f-L_n(f)\| \le C_m(1 + \| f\|)\left[(1-\rho_{m,n}) + \omega(f , (1-\rho_{m,n})^{\frac{1}{2}})\right] \qquad (3.12.3)$$

In addition, if f is differentiable on $[-\pi,\pi]$ then

$$\| f-L_n(f)\| \le C_m(1 + \| f\| + \| f'\|)((1-\rho_{m,n}) + (1-\rho_{m,n})^{\frac{1}{2}}\omega(f' ,(1-\rho_{m,n})^{\frac{1}{2}})) \qquad (3.12.4)$$

Now suppose (L_n) is a sequence of positive convolution operators which is saturated with order $(1-\rho_{m,n})$ where m is the smallest positive integer satisfying (3.3.1). If f has period $\frac{2\pi}{m}$ and $f' \in \text{Lip } 1$ then by Corollary 3.1

$$\| f-L_n(f)\| = O(1-\rho_{m,n})$$

so that $f \in S(L_n)$. Our next theorem shows that all the functions in $S(L_n)$ are of period $\frac{2\pi}{m}$.

<u>Theorem 3.12. Let (L_n) be a sequence of positive convolution operators of the form (3.1.1) which is saturated. Furthermore, let m be the smallest positive integer for which (3.3.1) holds. If $f \in S(L_n)$, then f has period $\frac{2\pi}{m}$.</u>

<u>Proof.</u> For each $0 < k < m$

$$\overline{\lim_{n\to\infty}} \frac{1-\rho_{k,n}}{1-\rho_{m,n}} = \infty$$

Since otherwise, (L_n) would be saturated with order $(1-\rho_{k,n})$ and this would contradict our choice of m. We wish to show that for each

$k \neq 0 \pmod{m}$

$$\overline{\lim} \frac{1-\rho_{k,n}}{1-\rho_{m,n}} = \infty \tag{3.12.5}$$

To see this, let $k = \ell m + j$ with $\ell, j \in N$, $0 < j < m$. Consider the two functions

$$h_1(x) = \sin^2 \frac{mx}{2} + \sin^2 \frac{(\ell m+j)x}{2}$$

and

$$h_2(x) = \sin^2 \frac{jx}{2} \; .$$

If x_o is a zero of h_1 then there are integers n_1 and n_2 such that

$$mx_o = 2n_1\pi$$

$$(\ell m+j)x_o = 2n_2\pi \; .$$

So that

$$jx_o = 2n_2\pi - 2n_1\ell\pi = 2\pi(n_2 - n_1\ell) \; .$$

Therefore, x_o is a zero of h_2. Since all the zeros of h_1 and h_2 are double zeros, every zero of h_1 is a zero of h_2 with the same multiplicity. Thus, there is a constant $C > 0$ such that

$$Ch_1(x) \geq h_2(x) \qquad x \in [-\pi, \pi] .$$

Integrating, we have for $n \in N$

$$\frac{C}{\pi} \int_{-\pi}^{\pi} \sin^2 \frac{mx}{2} \, d\mu_n(x) + \frac{C}{\pi} \int_{-\pi}^{\pi} \sin^2 \frac{(lm+j)x}{2} \, d\mu_n(x)$$

$$\geq \frac{1}{\pi} \int_{-\pi}^{\pi} \sin^2 \frac{jx}{2} \, d\mu_n(x)$$

This gives

$$\overline{\lim} \frac{1-\rho_{k,n}}{1-\rho_{m,n}} \geq \overline{\lim} \frac{1-\rho_{j,n}}{1-\rho_{m,n}} = \infty$$

which gives us (3.12.5)

Now suppose $f \in S(L_n)$. Then

$$\| f-L_n(f) \| = O(1-\rho_{m,n}) \qquad (n \to \infty)$$

If $k \neq 0 \pmod m$, then by virtue of (3.12.5), we must have $\hat{f}(k) = 0$. This shows that f has period $\frac{2\pi}{m}$.

We can use Theorem 3.12 to obtain saturation theorems which characterize the saturation class when the saturation order is $(1-\rho_{m,n})$. The following theorem should be compared with Theorem 3.2 and Sunouchi's Theorem (Theorem 3.3).

Theorem 3.13. *Let* (L_n) *be a sequence of positive convolution operators for which*

$$\lim_{n\to\infty} \frac{1-\rho_{k,n}}{1-\rho_{m,n}} = \Psi_k > 0 \qquad k = 1,2,\ldots \tag{3.12.6}$$

with m the smallest integer such that (3.12.6) holds. Then (L_n) is saturated with order $(1-\rho_{m,n})$. If $f \in S(L_n)$ then f has period $\frac{2\pi}{m}$ and

$$\sum_{k=1}^{\infty} \Psi_{km} A_{km}(f,x) \in L_\infty \tag{3.12.7}$$

If in addition $\left(\frac{1-\rho_{km,n}}{\Psi_{km}(1-\rho_{m,n})}\right)_{k=1}^{\infty} \in (C^*,C^*)$ with norms uniformly bounded in n, then $S(L_n) = \{f : f$ has period $\frac{2\pi}{m}$ and (3.12.7) holds$\}$.

The analogous version of Tureckii's Theorem (Theorem 3.6) is as follows.

Theorem 3.14. If (L_n) is a sequence of positive convolution operators such that

$$\lim_{n\to\infty} \frac{1-\rho_{mk,n}}{1-\rho_{m,n}} = k^2 \qquad k \in N \tag{3.12.8}$$

and

$$\overline{\lim_{n\to\infty}} \frac{1-\rho_{j,n}}{1-\rho_{m,n}} = \infty \qquad 1 \le j < m \tag{3.12.9}$$

Then (L_n) is saturated with order $(1-\rho_{m,n})$ and saturation class $S(L_n) = \{f : f$ has period $\frac{2\pi}{m}$ and $f' \in \text{Lip } 1\}$.

In the last theorem, we could replace (3.12.8) by suitable analogues to the condition given in Theorem 3.8. (e.g. $\lim_{n\to\infty} \frac{1-\rho_{2m,n}}{1-\rho_{m,n}} = 4$). The assumption (3.12.9) indicates that the saturation order is determined by $(1-\rho_{m,n})$ and not by lower coefficients.

We can also restate Theroem 3.10.

Theorem 3.15. Let (L_n) be a sequence of positive convolution operators and suppose there is a constant $C > 0$ with the property that for each $k \in N$, there is an $N(k)$ for which

$$\frac{1-\rho_{km,n}}{1-\rho_{m,n}} \geq Ck^2 \qquad n \geq N(k) \tag{3.12.10}$$

and

$$\overline{\lim_{n\to\infty}} \frac{1-\rho_{j,n}}{1-\rho_{m,n}} = \infty \qquad 1 \leq j < m \tag{3.12.11}$$

Then (L_n) is saturated with order $(1-\rho_{m,n})$ and saturation class $S(L_n) = \{f : f$ has period $\frac{2\pi}{m}$, $f' \in \text{Lip } 1\}$.

Here (3.12.10) could be replaced by the equivalent condition

3.12.12. There is a constant $C' > 0$ with the property that for each set $\mathcal{O}$ which contains the zeros of $\sin^2 \frac{mt}{2}$, there is a $N(\mathcal{O})$ such that

$$\int_{\mathcal{O}} \sin^2 \frac{mt}{2}\, d\mu_n(t) \geq C'(1-\rho_{m,n}) \qquad n > N(\mathcal{O})$$

The proofs of these last three theorems are the same as their version for the saturation order $(1-\rho_{1,n})$ except that Theorem 3.12 is used to guarantee that each function in the saturation class has period $\frac{2\pi}{m}$. To avoid repetition, we omit these proofs.

3.13. Notes.

3.13.1. It is sometimes useful to visualize the saturation theorems of this chapter in a distributional sense. Let $L_n(f) = f*d\mu_n$ be positive convolution operators, saturated, say with order $(1-\rho_{1,n})$. If $d\Psi_n = (1-\rho_{1,n})^{-1}(d\rho_o - d\mu_n)$, then for each $f \in C^{*(2)}$

$$\| f*d\Psi_n \| \leq C(\| f' \| + \| f'' \|)$$

where C does not depend on f. If we think of the measures $d\Psi_n$ as distributions acting on the space $C^{*(2)}$, their norms forma bounded sequence. Hence, there is a subsequence $(d\Psi_{n_k})$ which converges in the dual of $C^{*(2)}$ (under the weak topology induced by $C^{*(2)}$) to a distribution D which is tempered and of order at most 2. The "o" saturation theorem is merely a statement that $\hat{D}(k) \neq o$, if $k \neq o$. Since we can always work with functions whose mean value is 0, we can consider D to have an inverse D^{-1}. Here, $\widehat{D^{-1}}(k) = (\hat{D}(k))^{-1}$, $k \neq o$.

Tureckii's Theorem (Theorem 3.6) handles the case when $D = D^{(2)} = \left.\frac{d^2}{dx^2}\right|_o$ the second derivative operator evaluated at 0. The Sunouchi-Watari Theorem (Theorem 3.2) states that if $f \in S(L_n)$ (i.e. $\| f*d\Psi_n \| \leq M$, $n \in N$) then the distribution $f*D$ is a function in L_∞. The hypotheses of the direct theorem of Sunouchi (Theorem 3.3) say that the distributions $D^{-1}*d\Psi_n$ are measures with norms uniformly bounded in n. Hence, if $f*D \in L_\infty$

then

$$\| f*d\Psi_n \| = \| (f*D) * (D^{-1}*d\Psi_n) \| = O(1) .$$

The hypothesis of Theorem 3.10 essentially guarantee that $D = \alpha D_0^{(2)} + D_1 + D_2$, where $\| D_1 \|_2 < |\alpha|$ (here $\| \cdot \|_2$ means the norm of the distribution considered as a functional on $C^{*(2)}$) and D_2 is a measure which vanishes on a neighborhood of the origin. For more detailed discussions of distribution theory in saturation see E. Görlich$_1$ and J. Boman$_1$.

3.13.2. Most of the saturation theorems of this chapter have analogues on the line (and $R^{(n)}$). There is no general way to deduce results on the line from results on the circle (or vice versa). However, there is a technique (see Y. Katznelson$_1$(p.128)) for taking a function f (or measure) on R and deriving from it a function f_T on the circle with $\hat{f}_T(n) = \hat{f}(n)$, $n = 0, \pm 1, \ldots$ Also if $d\mu$ is non-negative on R then the corresponding measure $d\mu_T$ is non-negative.

Using this correspondence, it can be shown (see G. Sunouchi$_3$) that a function g defined and continuous on R is a Fourier-Stieljes transform of a measure $d\mu$ on R, if and only if for each $n \in N$ the sequence $(g(\frac{k}{n}))_{k=-\infty}^{\infty}$ is the sequence of Fourier-Stieljes coefficeints of a measure $d\mu_n$ on the circle, with $\int_{-\pi}^{\pi} |d\mu_n| = O(1)$. Positivity is also preserved (i.e. $d\mu$ is positive if and only if each $d\mu_n$ is positive).

Sunouchi has used these ideas to give examples where the direct theorem does not hold. He shows that if $0 < \alpha$ then there is a function $h \in L_1(R)$ such that $\lim_{t\to 0} |t|^{-\alpha} (1 - 2h(t)) = 1$, but $|t|^{-\alpha}(1 - 2h(t))$ is not a Fourier-Stieljes transforms. Therefore, the sequence of the measures $(d\mu_n)$ on the circle whose Fourier coefficients are $(h(\frac{k}{n}))_{k=-\infty}^{\infty}$ provide an example where

$\int_{-\pi}^{\pi} |d\mu_n| \leq M$ and

$$\lim_{n\to\infty} \frac{1-\rho_{k,n}}{n^{\alpha}} = k^{\alpha}$$

but the direct theorem does not hold.

The function h in the example is not positive definite. This leads to the following question which was communicated by Sunouchi . If $0 < \alpha < 2$ and h is positive definite (i.e. the Fourier transform of a non-negative function in L_1) such that $|t|^{-\alpha}(1 - 2h(t))$ is continuous at 0 then does it follow that $|t|^{-\alpha}(1 - 2h(t))$ is a Fourier-Stieljes transform. When $\alpha = 2$, this is true (see Theorem 3.6) .

3.13.3. Let $L_n(f) = f*d\mu_n$, with $(d\mu_n)$ satisfying the hypothesis of Theorem 3.2. It is important to point out that if $(\Psi_k^{-1}) \in (C^*,C^*)$, then the multiplier condition (3.5.3) is necessary as well as sufficient for the saturation class of (L_n) to be characterized by $S(L_n) = \{f : \sum_{-\infty}^{\infty} \Psi_k \hat{f}(k)e^{ikx} \in L_\infty\}$.

Suppose that (3.5.3) does not hold. Let $d\gamma_n$ be the Borel measure with mean value 0 , whose Fourier coefficeints are

$$2\hat{\gamma}_n(k) = \frac{1-\rho_{k,n}}{\Psi_k(1-\rho_{1,n})} \qquad k = \pm 1, \pm 2, \ldots$$

Then, $\int_{-\pi}^{\pi} |d\gamma_{n_j}| \to \infty$, for some subsequence (n_j) . Let $g \in C^*$ for which $\| g*d\gamma_{n_j}\| \to \infty$. The function $f \in C^*$ whose Fourier coefficients are $\Psi_k^{-1} \hat{g}(k)$ will satisfy $\sum_{-\infty}^{\infty} \Psi_k \hat{f}(k)e^{ikx} \in L_\infty$, but $f \notin S(L_n)$ (see arguments in proof of

Theorem 3.7) .

3.13.4. The transform technique gives a characterization of the saturation class in terms of Fourier transforms, while we would rather have a characterization in terms of the structural properties of f. This can be accomplished in the classical cases, since then the saturation class is usually

$\{\sum |k|^\alpha \hat{f}(k)e^{ikx} \in L_\infty\}$ for some $\alpha > 0$.

However, the structural characterization of these latter classes (see Butzer and Nessel_1) are themselves given in terms of approximation by linear operators. The simplest case being where $\alpha = 2$ and then

$\sum_{-\infty}^{\infty} k^2 \hat{f}(k)e^{ikx} \in L_\infty$ if and only if

$$\| f(x+t) + f(x-t) - 2f(x) \| = O(t^2)$$

which is of course a statement about the degree of approximation of f by the operators L_t, $L_t(f,x) = \frac{1}{2}(f(x+t) + f(x-t))$.

Thus, when we seek a characterization of the saturation class in terms of structural properties of the function we are really asking for a statement comparing the approximation of the original sequence of operators, with a statement about the approximation by a new, hopefully simple sequence of operators. H.S. Shapiro_1 has made a penetrating study into the comparison of linear methods of approximation given by convolution with dilates of a measure. In this kind of development there is a unified setting for direct and inverse theorems since both are comparison theorems.

3.13.5. We gave in Section 3.10 an example of a sequence of positive linear operators which under the usual definition of saturation is saturated with respect to two different orders, which in turn gave two distinct saturation classes. We can give a slightly more complicated example which has some other interesting features.

Define (L_n) by $L_n = J_n$ the Jackson operator if $n \neq 2^{4^k}$ and $L_n = \sigma_n$ the Fejer operator if $n = 2^{4^k}$. The sequence (L_n) is saturated (usual definition) with order $\phi_n = \frac{1}{n}$. Since, if $\| f - L_n(f) \| = o(\phi_n)$, then $\| f - \sigma_{\lambda_k}(f) \| = o(\lambda_k^{-1})$, $\lambda_k = 2^{4^k}$ which implies that f is constant. With this saturation order, the saturation class $S(L_n)$ contains every function f for which $f \in \text{Lip } 1$ and $\tilde{f} \in \text{Lip } 1$ and thus, in particular, all $f \in \text{Lip}^* \alpha$, $\alpha > 1$.

We can construct another sequence of numbers (Ψ_n), in such a way that (L_n) is saturated with order (Ψ_n) but has a different saturation class than for (ϕ_n). Namely, define Ψ_n as $2^{-2^{2k+2}}$ when $2^{2^{2k+1}} \leq n \leq 2^{2^{2k+2}}$, $k = 0,1,\ldots$, For $2^{2^{2k}} \leq n \leq 2^{2^{2k+1}}$, we can define Ψ_n in such a way that it decreases monotonically from the value $2^{-2^{2k}}$ for $n = 2^{2^{2k}}$ to the value $2^{-2^{2k+2}}$ for $n = 2^{2^{2k+1}}$. We can also require that the ratios $\frac{\Psi_n}{\Psi_{n+1}}$ are bounded uniformly in n. To see the latter, we just note that

$2^{2^{2k+2}} \cdot 2^{-2^{2k}} = 8^{2^{2k}}$ and we have more than $2^{2^{2k}}$ steps available to us. Thus, we can at least achieve this with ratios ≤ 8, for example.

The saturation class for (Ψ_n) is now $\{f : f' \in \text{Lip } 1\}$. Since, if f is in the saturation class then $\| f - J_{\lambda_k} \| = O(\lambda_k^{-2})$; $\lambda_k = 2^{2^{2k+1}}$.

This saturation class is therefore different from the saturation class when we use the order (ϕ_n). The sequences (ϕ_n) and (Ψ_n) are both monotonically decreasing and have bounded ratios (i.e. $\frac{\phi_n}{\phi_{n+1}}$ and $\frac{\psi_n}{\psi_{n+1}}$ are bounded) . The significance of this latter condition will be seen more clearly when we disuss inverse theorems in Chapter 8 . If we use our definition for saturation, then the example (L_n) just constructed is saturated with order (δ_n) where $\delta_n = n^{-2}$ $n \neq 2^{4^k}$ and $\delta_n = n^{-1}$, $n = 2^{4^k}$. The saturation is now uniquely determined and is $\{f : f' \in \text{Lip } 1\} = \text{Lip}^* \, 2$.

CHAPTER 4

TRIGONOMETRIC POLYNOMIAL OPERATORS

4.1. Introduction. We wish to now consider operators with ranges contained in the space of trigonometric polynomials. We will say (L_n) is a sequence of trigonometric polynomial operators if there exists integers $\alpha \geq 1$ and $\beta \geq 0$, such that for each $n \in N$, L_n maps C^* into $T_{\alpha n+\beta}$. Of course, we have already seen several examples of this type. Our main concern will be to examine the limitations on the degree of approximation, by such a sequence of operators. In this chapter, (L_n) will always denote a sequence of positive trigonometric polynomial operators.

In order to estimate the degree of approximation of $(L_n(f))$ to a function $f \in C^*$, we can use our fundamental estimates developed in Chapter 2 (Theorem 2.4). Thus, if

$$\| g - L_n(g) \| = O(\lambda_n^2)$$

for the three functions $g = 1$, $\cos x$, $\sin x$, then for each function $f \in C^*$

$$\| f - L_n(f) \| = O(\lambda_n^2 + \omega(f,\lambda_n)) \qquad (4.1.1)$$

The faster (λ_n) tends to 0, the better the degree of approximation guaranteed by (4.1.1). But then, how fast can (λ_n) tend to 0? This is easily answered. In Chapter 1 (see discussion after Theorem 1.8,) we noted there is an $f \in \mathrm{Lip}\ 1$ for which $E_n^*(f) \geq \frac{1}{n}$, $n \in N$. Thus, we cannot have $\lambda_n = o(n^{-1})$. we have already seen examples of operators for which $\lambda_n = O(n^{-1})$, namely, the Jackson operators and the Jackson-Matsuoka operators

The argument given above establishes the following theorem of P.P. Korovkin$_4$.

Theorem 4.1. If (L_n) is a sequence of positive trigonometric polynomial operators, then for g one of the three functions 1, $\cos x$ or $\sin x$

$$\| g-L_n(g) \| \neq o(n^{-2}) \tag{4.1.2}$$

For convolution operators, we always have $L_n(1) = 1$. It is the functions $\sin x$ and $\cos x$ which are the essential limitations in (4.1.2) An optimal behavior occurs when

$$\| 1-L_n(1) \| = o(n^{-2}) \tag{4.1.3}$$

and

$$\| g-L_n(g) \| = O(n^{-2}) \tag{4.1.4}$$

for $g = \cos x$, $\sin x$. When (4.1.3) and (4.1.4) hold we say that (L_n) is optimal.

Such a sequence of operators is optimal in another sense. Consider those sequences (L_n) which satisfy (4.1.3) and give the Jackson estimates

$$\| f-L_n(f) \| = O(n^{-2} + n^{-1} \omega(f', n^{-1})) \tag{4.1.5}$$

for continuously differentiable f. They are exactly the optimal sequences. Theorem 2.4 provides the estimate (4.1.5) when (L_n) is optimal. That any sequence (L_n) for which (4.1.5) holds is optimal is clear, since the functions $\cos x$ and $\sin x$ have first derivatives in Lip 1 and thus must be approximated with order $O(n^{-2})$,

4.2. Examples of Optimal Sequences. The sequence (J_n) of Jackson operators introduced in Section 2.1 is optimal. So is the generalization, the sequence of the Jackson-Matsuoka operators $(J_{n,p,q})$, with $q \geq 2$. For a sequence (L_n) of convolution operators, $L_n(f) = f*T_n$, the condition of optimality reduces simply to having

$$1-\rho_{1,n} = \frac{2}{\pi}\int_{-\pi}^{\pi} \sin^2 \frac{t}{2}\, T_n(t)dt = O(n^{-2}).$$

That is to say, the first Fourier coefficients determine whether (L_n) is optimal.

This leads us to consider the following extremal problem: minimize $\frac{1}{\pi}\int_{-\pi}^{\pi}\sin^2 \frac{t}{2}\, T(t)dt$ over all non-negative trigonometric polynomials T of degree $\leq n$ with $\frac{1}{\pi}\int_{-\pi}^{\pi} T(t)dt = 1$. This extremal problem was solved by L. Fejér$_1$ while P.P. Korovkin$_3$ pointed out its importance in approximation theory.

Theorem 4.2. Let K_n be the trigonometric polynomial of degree n given by

$$K_n(t) = \frac{1}{n+2}\left(\frac{\sin\frac{\pi}{n+2}\cos\frac{(n+2)t}{2}}{\cos t - \cos\frac{\pi}{n+2}}\right)^2 \tag{4.2.1}$$

Then

$$\inf\left\{\frac{1}{\pi}\int_{-\pi}^{\pi}\sin^2\frac{t}{2}\, T(t)dt : T \in \mathcal{T}_n,\ T(t) \geq 0,\ \frac{1}{\pi}\int_{-\pi}^{\pi}T(t)dt = 1\right\}$$

$$= \frac{1}{\pi}\int_{-\pi}^{\pi}\sin^2\frac{t}{2}\, K_n(t)dt = \frac{1}{2}\left(1-\cos\frac{\pi}{n+2}\right) = \sin^2\frac{\pi}{2n+4}.$$

Remark: The operators $U_n(f) = f*K_n$ are called the Fejér-Korovkin operators and (U_n) is another example of an optimal sequence.

<u>Proof</u>. We use the quadrature formula (see Theorem 1.13) for trigonometric polynomials with nodes at

$$\frac{\pi}{(n+2)} + \frac{2k\pi}{(n+2)} , \quad k = -[\frac{n+3}{2}], \ldots, -1, 0, 1, \ldots, [\frac{n}{2}]$$

which is exact for polynomials of degree $\le n+1$. Note, that the nodes are the zeros of $\cos(\frac{n+2}{2}t)$ in $[-\pi,\pi)$. This quadrature formula gives immediately that $\frac{1}{\pi}\int_{-\pi}^{\pi} K_n(t)dt = 1$.

Let T be any non-negative trigonometric polynomial of degree $\le n$ with $\frac{1}{\pi}\int_{-\pi}^{\pi} T(t)dt = 1$. Since $\sin^2\frac{t}{2}T(t)$ is a polynomial of degree $\le n+1$, we have

$$\frac{1}{\pi}\int_{-\pi}^{\pi} \sin^2\frac{t}{2}\,T(t)dt = \frac{2}{n+2}\sum_{k=-[\frac{n+3}{2}]}^{[\frac{n}{2}]} \sin^2\frac{(2k+1)\pi}{2n+4}\; T\left(\frac{2k+1}{n+2}\pi\right)$$

$$\ge \left(\sin^2\frac{\pi}{2(n+2)}\right)\left(\frac{2}{n+2}\right)\sum_{k=-[\frac{n+3}{2}]}^{[\frac{n}{2}]} T\left(\frac{2k+1}{n+2}\pi\right) \ge \left(\sin^2\frac{\pi}{2n+4}\right)\frac{1}{\pi}\int_{-\pi}^{\pi} T(t)dt$$

$$= \sin^2\frac{\pi}{2n+4}$$

For K_n, we have $K_n\left(\frac{(2k+1)\pi}{n+2}\right) = 0$ if $k \ne 0, -1$, Therefore,

$$\frac{1}{\pi}\int_{-\pi}^{\pi} \sin^2\frac{t}{2}\,K_n(t)dt = \left(\sin^2\frac{\pi}{2n+4}\right)\left(\frac{2}{n+2}\right)\left(K_n\left(\frac{-\pi}{n+2}\right) + K_n\left(\frac{\pi}{n+2}\right)\right)$$

$$= \left(\sin^2\frac{\pi}{2n+4}\right)\left(\frac{1}{\pi}\int_{-\pi}^{\pi} K_n(t)dt\right) = \sin^2\frac{\pi}{2n+4} .$$

A general method for constructing optimal sequences of operators was given by Korovkin$_6$ (see also V. Baskakov$_1$) Let ϕ be a continuous function on

$[0,1]$ such that for each $n \in N$

$$C_n = \sum_{k=0}^{n} \phi^2(\tfrac{k}{n}) > 0$$

Then, the kernel

$$T_n(x) = \frac{1}{2} C_n^{-1} \left| \sum_{k=0}^{n} \phi(\tfrac{k}{n}) e^{ikt} \right|^2 \tag{4.2.2}$$

is a non-negative trigonometric polynomial of degree $\leq n$. Under certain conditions on ϕ the operators generated by convolution with T_n are optimal.

Theorem 4.3. *If for some* $M > 0$, $\phi \in \mathrm{Lip}(1,M)$ *on* $[0,1]$ *with* $\int_0^1 \phi^2(t)dt > 0$ *and* $\phi(0) = \phi(1) = 0$, *then the sequence of operators* (L_n) *given by* $L_n(f) = f * T_n$ *with* T_n *as in* (4.2.2) *is optimal.*

Remark. It may be necessary to take n suitably large to guarantee that $C_n > 0$.

Proof. Since $T_n(x) = \frac{1}{2} C_n^{-1} \left(\sum_{k=0}^{n} \phi(\frac{k}{n}) e^{ikt} \right) \left(\sum_{k=0}^{n} \phi(\frac{k}{n}) e^{-ikt} \right)$ we have

$$\rho_{1,n} = C_n^{-1} \sum_{k=0}^{n-1} \phi(\tfrac{k}{n})\, \phi(\tfrac{k+1}{n})$$

and therefore

$$n^2(1-\rho_{1,n}) = C_n^{-1} n^2 \sum_{k=0}^{n-1} [\phi^2(\tfrac{k}{n}) - \phi(\tfrac{k}{n})\, \phi(\tfrac{k+1}{n})]$$

$$= C_n^{-1} n^2 \sum_{k=0}^{n-1} \tfrac{1}{2}(\phi(\tfrac{k}{n}) - \phi(\tfrac{k+1}{n}))^2 \tag{4.2.3}$$

We have used the fact that $\phi(0) = \phi(1) = 0$ in manipulating the sums.

Since ϕ is in Lip(1,M)

$$|\phi(\tfrac{k}{n}) - \phi(\tfrac{k+1}{n})| \leq \frac{M}{n} \tag{4.2.4}$$

Using this in (4.2.3) shows that

$$n^2(1-\rho_{1,n}) \leq \frac{1}{2} M^2 n \, C_n^{-1} = \frac{1}{2} M^2 \left(\sum_{k=0}^{n} \frac{1}{n} \phi^2(\tfrac{k}{n})\right)^{-1} \tag{4.2.5}$$

Realizing that the sum on the right side of (4.2.5) is a Riemann sum for $\int_0^1 \phi^2(t)dt$, we can find a constant $C > 0$ such that for $n \in N$

$$n^2(1-\rho_{1,n}) \leq C\left(\int_0^1 \phi^2(t)dt\right)^{-1} \tag{4.2.6}$$

Of course, (4.2.6) gives that (L_n) is optimal.

We should note here, that when $\phi(t) = 1-2|t-\frac{1}{2}|$ the construction of Korovkin recovers the Jackson operators and when $\phi(t) = \sin \pi t$, it recovers the Fejér-Korovkin operators.

4.3. Saturation of Optimal Sequences. When (L_n) is optimal, we have seen that for one of the two functions $g = \cos x$ or $\sin x$

$$\| g-L_n(g)\| \neq o(n^{-2}) .$$

The question arises whether there can be a non-constant function f_o for which $\| f_o - L_n(f_o)\| = o(n^{-2})$. Or what is the same thing, is every optimal sequence saturated with order (n^{-2}). This question is not settled in the general case although when each L_n is a convolution operator the question is answered by means of the following lemma .

Lemma 4.1. *There is a constant* $C_0 > 0$ *such that for any integers* n *and* k *with* $n \geq k \geq 1$ *we have*

$$\frac{1}{\pi} \int_{-\pi}^{\pi} \sin^2 \frac{kt}{2} T_n(t)dt \geq C_0 \frac{k^2}{n^2} \tag{4.3.1}$$

whenever T_n *is a non-negative trigonometric polynomial of degree* n *with*

$$\frac{1}{\pi} \int_{-\pi}^{\pi} T_n(t)dt = 1$$

Proof. The proof is based on the fact (see remarks following Theorem 1.8) that the function $h(x) = |\sin x|$ has degree of approximation $E_n^*(|\sin x|) \geq \frac{1}{4\pi n}$

We let T_n^* denote the polynomial in T_n for which

$$\| h - T_n^* \| = E_n^*(h) \qquad n = 1,2, \ldots$$

Then $h - T_n^*$ alternately assumes its maximum and minimum at least $2n+2$ distinct points in $[-\pi,\pi)$. This is from the Chebyshev alternation theorem (Theorem 1.1)

If $k \in N$, let $h_k(x) = h(kx)$ and $T_{n,k}^*(x) = T_n^*(kx)$ then $h_k - T_{n,k}^*$ alternately assumes maximums and minimums at, at least $(2n+2)k \geq 2nk+2$ points in $[-\pi,\pi)$. Thus, $T_{n,k}^*$ is the best uniform approximation to h_k of degree $\leq nk$. Therefore, if $k , n \in N$,

$$E_{nk}^* (|\sin kx|) \geq \frac{1}{4\pi n} .$$

Now if m is any integer $\geq k$ choose $n \geq 1$ so that $nk \leq m < (n+1)k$ then

$$E^*_m(|\sin kx|) \geq E^*_{k(n+1)}(|\sin kx|) \geq \frac{1}{4\pi(n+1)} = \frac{k}{4\pi kn} \cdot \frac{n}{n+1}$$

$$\geq \frac{1}{2} \frac{k}{4\pi kn} \geq \frac{k}{8\pi m}$$

We can now show that (4.3.1) holds with $C_o = (16\pi)^{-2}$. For each $n \in N$, $h_k * T_n$ is a trigonometric polynomial of degree n, so that,

$$\| h_k - h_k * T_n \| \geq \frac{k}{8\pi n} \quad \text{when } n \geq k .$$

Therefore,

$$\frac{k}{8\pi n} \leq \| h_k - h_k * T_n \| = \| \frac{1}{\pi} \int_{-\pi}^{\pi} (|\sin kt| - |\sin kx|) T_n(x-t)dt \|$$

$$\leq \| \frac{1}{\pi} \int_{-\pi}^{\pi} |\sin kt - \sin kx| T_n(x-t)dt \|$$

$$= \| \frac{2}{\pi} \int_{-\pi}^{\pi} |\cos k \frac{(x+t)}{2} \sin k \frac{(x-t)}{2}| T_n(x-t)dt \|$$

$$\leq \frac{2}{\pi} \int_{-\pi}^{\pi} | \sin \frac{kt}{2} | T_n(t)dt$$

Using the Cauchy-Schwarz inequality for positive functionals, we find

$$\frac{k}{16\pi n} \leq \frac{1}{\pi} \int_{-\pi}^{\pi} |\sin \frac{kt}{2}| T_n(t)dt \leq \left(\frac{1}{\pi} \int_{-\pi}^{\pi} T_n(t)dt\right)^{\frac{1}{2}} \left(\frac{1}{\pi} \int_{-\pi}^{\pi} \sin^2 \frac{kt}{2} T_n(t)dt\right)^{\frac{1}{2}}$$

and thus

$$\left(\frac{k}{16\pi n}\right)^2 \leq \frac{1}{\pi}\int_{-\pi}^{\pi} \sin^2 \frac{kt}{2} T_n(t)dt$$

which establishes the lemma.

The following theorem of P.C. Curtis[1] is in the same spirit as a "o" saturation theorem and in particular when (L_n) is optimal it shows (L_n) is saturated with order (n^{-2}) .

Theorem 4.4. *Let* (L_n) *be a sequence of positive trigonometric convolution operators,* $L_n(f) = f*T_n$, $T_n \geq 0$ *and even. If* $f \in C^*$

$$\lim n^{-2}\|f-L_n(f)\| = 0 \qquad (4.3.2)$$

then f *is constant.*

Proof. The proof is the transform technique. If f satisfies (4.3.2) then

$$\lim_{n\to\infty} n^2(\hat{f}(k)(1-2\hat{T}_n(k))) = 0$$

From Lemma 4.1, $n^2(1-2\hat{T}_n(k))$ is bounded away from 0 , thus we must have $\hat{f}(k) = 0, k = \pm 1, \pm 2, \ldots$. Therefore, f is constant.

The following saturation theorem of DeVóre[3] characterizes the saturation class for any sequence of optimal convolution operators.

Theorem 4.5. *Let* (L_n) *be an optimal sequence of convolution operators* , $L_n(f) = f*T_n$. *Then* $S(L_n) = \mathrm{Lip}^* 2$.

Proof. Let α and β be the integers for which $L_n : C^* \to T_{\alpha n+\beta}$. From Lemma 4.1 we have for $n \geq k$

$$\frac{1}{\pi} \int_{\pi}^{\pi} \sin^2 \frac{kt}{2} T_n(t)dt \geq \frac{C_0 k^2}{(\alpha n+\beta)^2} \geq C \frac{k^2}{n^2}$$

for some constant C independent of k and n .

Also, from the definition of optimal (4.1.4)

$$\frac{1}{\pi} \int_{-\pi}^{\pi} \sin^2 \frac{t}{2} T_n(t)dt = O(n^{-2})$$

So that condition (3.11.2) is satisfied. The theorem then follows from Theorem 3.10.

An immediate application of Theorem 4.3 gives the saturation of the Fejèr-Korovkin operators (see P.L. Butzer and E. Görlich$_1$). We could have also derived their saturation properties directly from Theorem 3.6, by first evaluating the $\lim_{n\to\infty} n^2(1-\rho_{k,n})$ which in this case exists.

Corollary 4.1. The sequence of Fejér-Korovkin operators (U_n) is saturated with order (n^{-2}) and saturation class $S(U_n) = \text{Lip}^* \, 2$,

Theorem 4.5 also gives the following corollary which gives the saturation for the operators introduced in Theorem 4.3.

Corollary 4.2. If the conditions of Theorem 4.3 are satisfied then the operators (L_n) are saturated with order (n^{-2}) and saturation class $S(L_n) = \text{Lip}^* \, 2$

<u>4.4. Quasi-optimal Sequences</u>. In the definition of an optimal sequence, we have required that

$$\| g-L_n(g) \| = O(n^{-2})$$

for the two functions $g = \cos x$, $\sin x$. Let us weaken this assumption to just requiring that there is a non-constant function $f_o \in C^*$ such that

$$\| f_o-L_n(f_o) \| = O(n^{-2}) \qquad (4.4.1)$$

A sequence of operators which satisfies (4.4.1) will be called quasi-optimal. We still require that $\|1-L_n(1)\| = o(n^{-2})$.

It is easy to give an example of sequence of operators (L_n) which is quasi-optimal but not optimal. The idea is to let the second Fourier coefficients determine the approximation properties of (L_n). The example is based on the same ideas as our construction of the operators $\bar{A}_n$ from A_n in Sections 3.1 and 3.2. Let K_n be the Jackson kernel and consider the modified kernels $\bar{K}_n(t) = (1-n^{-1})K_n(t) + \frac{1}{2} n^{-1}[K_n(t-\pi) + K_n(t+\pi)]$. Then the sequence of operators $(\bar{L}_n)$ with $\bar{L}_n(f) = f*\bar{K}_n$ is quasi-optimal but not optimal. We have

$$\frac{1}{\pi} \int_{-\pi}^{\pi} \sin^2 \frac{t}{2} \bar{K}_n(t) \geq \frac{1}{2n\pi} \int_{-\pi}^{\pi} \sin^2 \frac{t}{2} K_n(t-\pi)dt$$

and $\frac{1}{\pi} \int_{-\pi}^{\pi} \sin^2 \frac{t}{2} K_n(t-\pi)dt$ converges to $\sin^2 \frac{\pi}{2} = 1$. This shows $(\bar{L}_n)$ is not optimal. To show that $(\bar{L}_n)$ is quasi-optimal we need only consider the function $f_o(t) = \cos 2t$, for then

$$\| f_o(x) - \bar{L}_n(f_o,x) \|$$

$$= \| \frac{1}{\pi} \int_{-\pi}^{\pi} (\cos 2x - \cos 2t)\, \bar{K}_n(t-x)dt \| \leq \frac{2}{\pi} \int_{-\pi}^{\pi} \sin^2 t\, \bar{K}_n(t)dt \; .$$

Now,

$$\int_{-\pi}^{\pi} \sin^2 t\, K_n(t-\pi)dt - \sin^2 \pi = O(n^{-2}) \; ,$$

since $\sin^2 t$ has a first derivative in Lip 1 . Similarly

$$\int_{-\pi}^{\pi} \sin^2 t\, K_n(t+\pi)dt = O(n^{-2})$$

Thus

$$\frac{2}{\pi} \int_{-\pi}^{\pi} \sin^2 t\, \bar{K}_n(t)dt = \frac{2}{\pi} \int_{-\pi}^{\pi} \sin^2 t\, K_n(t) + O(n^{-2}) = O(n^{-2})$$

and this shows that $(\bar{L}_n)$ is quasi-optimal.

As for optimal operators, we cannot in general determine the saturation properties of quasi-optimal operators. However, when in addition each L_n is a convolution operator the following theorem (DeVore_3) determines the saturation of (L_n)

Theorem 4.6. Let (L_n) be a quasi-optimal sequence of convolution operators $L_n(f) = f*T_n$, $T_n \geq 0$ and even. Then there is a positive integer m for which

$$\frac{1}{\pi} \int_{-\pi}^{\pi} \sin^2 \frac{mt}{2}\, T_n(t)dt = O(n^{-2}) \qquad (4.4.2)$$

(L_n) is saturated with order (n^{-2}) and saturation class $S(L_n) = \{f : f$ has period $2\pi m_0^{-1}$, $f \in \text{Lip}^* 2$ } , <u>where</u> m_0 <u>is the smallest positive integer for which</u> (4.4.2) <u>holds</u>.

Proof. Let f_0 be a non-constant function in C^* such that

$$\| f_0 - L_n(f_0) \| = O(n^{-2})$$

Then for some $m \in N$, $\hat{f}_0(m) \neq 0$. Evaluating the m-th Fourier coefficients of $f_0 - L_n(f_0)$ we find

$$\hat{f}_0(m)(1 - 2\hat{T}_n(m)) = O(n^{-2})$$

and thus

$$1 - 2\hat{T}_n(m) = O(n^{-2})$$

which establishes (4.4.2) . The remainder of the theorem follows from Theorem 3.15 , since condition (3.12.10) is clearly satisfied via Lemma 4.1.

4.5. Non-Optimal Sequences. We have seen that the operators given by (4.2.2) are optimal when ϕ is in Lip 1 on [0,1] and $\phi(0) = \phi(1) = 0$. When ϕ does not vanish at the end points, then these operators are no longer optimal, but under certain restrictions on ϕ we can still determine their saturation properties.

Theorem 4.7. <u>Let</u> ϕ <u>be in</u> $\text{Lip}^* 2$ <u>on</u> [0,1] . <u>If</u> $\phi^2(0) + \phi^2(1) > 0$, <u>then the sequence of operators</u> (L_n) <u>determined by convolution with the trigonometric polynomials</u> T_n <u>given in</u> (4.2.2) <u>is saturated with order</u> (n^{-1}) <u>and saturation class</u> $S(L_n) = \{ f : \tilde{f} \in \text{Lip } 1 \}$.

Proof. For the real Fouriercoefficients $\rho_{k,n}$ of T_n, we have

$$\rho_{k,n} = C_n^{-1} \sum_{s=0}^{n-k} \phi(\frac{s}{n})\, \phi(\frac{s+k}{n}) \qquad k = 0,1, \ldots, n$$

Therefore,

$$n(1-\rho_{k,n}) = nC_n^{-1} \left(\sum_{s=0}^{n} \phi^2(\frac{s}{n}) - \sum_{s=0}^{n-k} \phi(\frac{s}{n})\phi(\frac{s+k}{n})\right)$$

$$= nC_n^{-1}\left[\frac{1}{2} \sum_{s=0}^{n-k} (\phi(\frac{s}{n}) - \phi(\frac{s+k}{n}))^2 + \frac{1}{2} \sum_{s=0}^{k-1} (\phi^2(\frac{s}{n}) + \phi^2(1 - \frac{s}{n})\right] \qquad (4.5.1)$$

Now, ϕ is in $\mathrm{Lip}(1,M_1)$ for some constant $M_1 > 0$. Hence, the first term on the right hand side of (4.5.1) can be estimated as

$$nC_n^{-1} \sum_{s=0}^{n-k} (\phi(\frac{s+k}{n}) - \phi(\frac{s}{n}))^2 \le nC_n^{-1} \sum_{s=0}^{n} M_1^2 n^{-2} = O(C_n^{-1}) = o(1)$$

For the second term on the right hand side of (4.5.1), we have

$$\lim_{n\to\infty} nC_n^{-1} = \lim_{n\to\infty} (n^{-1} \sum_{s=0}^{n} \phi^2(\frac{s}{n}))^{-1} = (\int_0^1 \phi^2(t)dt)^{-1} \qquad (4.5.2)$$

So, from the continuity of ϕ, we find

$$\lim_{n\to\infty} \frac{1}{2} nC_n^{-1} \sum_{s=0}^{k-1} (\phi^2(\frac{s}{n}) + \phi^2(1-\frac{s}{n})) = \frac{1}{2} k(\phi^2(0) + \phi^2(1))(\int_0^1 \phi^2(t)dt)^{-1}.$$

We have just shown that

$$\lim_{n\to\infty} n(1-\rho_{k,n}) = \frac{1}{2} k(\phi^2(0) + \phi^2(1))(\int_0^1 \phi^2(t)dt)^{-1}$$

which shows by virtue of the Sunouchi-Watari theorem (Theorem 3.2) that (L_n) is saturated with order (n^{-1}) and $S(L_n) \subseteq \{ f . \tilde{f} \in \text{Lip } 1\}$.

We wish now to show that if $\tilde{f} \in \text{Lip } 1$ then $f \in S(L_n)$. If we apply Abel summation by parts twice, we find

$$\frac{1}{2} \rho_{o,n} + \sum_{k=1}^{n} \rho_{k,n} \cos kt = \sum_{k=0}^{n-2} (k+1) \Delta^2 \rho_{k,n} F_k(t) + n\Delta \rho_{n-1,n} F_{n-1}(t)$$

$$+ \rho_{n,n} D_n(t)$$

where F_k is the Fejér kernel and D_n is the Dirichlet kernel. Therefore,

$$f - L_n(f) = \sum_{k=0}^{n-2} (k+1) \Delta^2 \rho_{k,n} (f - \sigma_k(f)) + n\Delta\rho_{n-1,n} (f - \sigma_{n-1})(f)) +$$

$$+ \rho_{n,n} (f - S_n(f)) \qquad (4.5.3)$$

For the last term in (4.5.3), we have $\rho_{n,n} = O(C_n^{-1})$, so that

$$|\rho_{n,n}| \; \| f - S_n(f) \| = O(C_n^{-1}) = O(n^{-1}) \qquad (4.5.4)$$

For the second term on the right hand side of (4.5.3), we find from the formula for $\rho_{k,n}$ that $n|\Delta\rho_{n-1,n}| = O(n\ C_n^{-1}) = O(1)$. Now, $\| f - \sigma_n(f) \| = O(n^{-1})$, since $\tilde{f} \in \text{Lip } 1$ (see Theorem 3.4) Thus,

$$\| n\Delta\rho_{n-1,n} (f - \sigma_n(f) \| = O(|\Delta\rho_{n-1,n}|) = O(C_n^{-1}) = O(n^{-1}) \qquad (4.5.5)$$

Finally, in view of (4.5.4) and (4.5.5), it is enough to show that the first term on the right hand side of (4.5.3) is $O(n^{-1})$. Since $\| f - \sigma_k(f) \| < Ck^{-1}$, for some constant C not depending on k , we need only show that

$$\sum_{k=0}^{n-2} |\Delta^2 \rho_{k,n}| = O(n^{-1}) \qquad (4.5.6)$$

To see that (4.5.6) holds, we write

$$\Delta^2 \rho_{k,n} = \rho_{k,n} - 2\rho_{k+1,n} + \rho_{k+2,n}$$

$$= C_n^{-1} \sum_{s=0}^{n-k-2} \phi(\tfrac{s}{n}) \left(\phi(\tfrac{s+k}{n}) - 2\phi(\tfrac{s+k+1}{n}) + \phi(\tfrac{s+k+2}{n})\right)$$

$$+ C_n^{-1} [\phi(\tfrac{n-k-1}{n})\phi(\tfrac{n-1}{n}) + \phi(\tfrac{n-k}{n})\phi(1) - 2\phi(\tfrac{n-k-1}{n}) \phi(1)] \qquad (4.5.7)$$

For the first sum on the right hand side of (4.5.7), we use that fact that $\phi \in \mathrm{Lip}^*(2,M_2)$ for some M_2 and thus

$$|C_n^{-1} \sum_{s=0}^{n-k-2} \phi(\tfrac{s}{n}) (\phi(\tfrac{s+k}{n}) - 2\phi(\tfrac{s+k+1}{n}) + \phi(\tfrac{s+k+2}{n}))|$$

$$\leq M_2^2 \cdot M\, C_n^{-1}\, n^{-1}$$

where $M = \| \phi \| [0,1]$.

For the second term on the right hand side of (4.5.7), we need only use the fact that $\phi \in \mathrm{Lip}(1,M_1)$ to find

$$C_n^{-1} \, |\phi(\tfrac{n-k-1}{n})(\phi(\tfrac{n-1}{n}) - \phi(1)) + \phi(1)(\phi(\tfrac{n-k}{n}) - \phi(\tfrac{n-k-1}{n}))|$$

$$\leq 2M \cdot M_1\, C_n^{-1}\, n^{-1}$$

These two inequalities combine to show that $|\Delta^2 \rho_{k,n}| \leq M'\, C_n^{-1}\, n^{-1}$ for some constant $M' > 0$, with M' independent of k and n . Hence

$$\sum_{0}^{n-2} |\Delta^2 \rho_{k,n}| \leq M'\, C_n^{-1} = 0(n^{-1})$$

as desired.

4.6. Notes.

4.6.1. As we have mentioned, there remains the open problem to show that if (L_n) is an optimal sequence of operators then $\| f - L_n(f) \| = o(n^{-2})$ implies f is constant. There is an anolgous definition of optimal sequence in the algebraic case which is given in Chapter 6. In the algebraic case, the corresponding problem has been settled. However, the techniques used in the algebraic case do not carry over to the trigonometric case. The study of optimal sequences was initiated in the papers of Görlich and $\text{Stark}_{1,2}$ (see also Görlich_2).

4.6.2. In Theorem 4.7., we avoided showing that $(k^{-1} n(1-\rho_{k,n}))_{k=1}^{\infty}$ are multipliers in (C^*,C^*) with norms uniformly bounded in n. However, our arguments were very similar to what would be used to show the multiplier condition. It would seem that if we had a more delicate test (than quasi-convexity) we would be able to weaken the hypothesis that $\phi \in \text{Lip}^* 2$, which is assumed in Theorem 4.7. Note that for the inverse part of Theorem 4.7, we only needed $\phi \in \text{Lip}\ 1$.

4.6.3. Lemma 4.1 can be deduced easily from the inequality

$$2\hat{T}_n(k) \leq \cos \pi([\tfrac{n}{k}] + 2)^{-1}$$

which holds (see G. Szego_1) whenever $T_n \geq 0$ and $\frac{1}{\pi}\int_{-\pi}^{\pi} T_n(t)dt = 1$. This inequality can be derived from the Fejer result $k = 1$ and it is the best possible upperbound for each k and n.

4.6.4. We have seen that positive trigonometric operators are limited in their efficiency of approximation because of the saturation phenomena. Thus, they can only give the Jackson estimates for smoothness up to $f' \in \mathrm{Lip}\ 1$.

If positivity of the kernel is dropped but replaced instead by the requirement that T_n is even with at most $2k$ changes of sign, k fixed, then the saturation phenomena can be delayed to order (n^{-2k-2}) and class $\{f : f^{(2k-1)} \in \mathrm{Lip}\ 1\}$. This is contained in the recently developed theory for approximation by convolution with kernels having a finite number of changes of sign . These investigations were initiated in the paper of P.L. Butzer, R.J. Nessel, and K. Scherer$_1$, who showed that if (L_n) was a sequence of operators given by convolution with T_n , T_n has at most $2k$ changes of sign, then for g one of the functions $1, \cos x, \ldots, \cos kx, \sin x, \ldots, \sin kx$, $\| g - L_n(g) \| \neq o(n^{-2k-2})$. Subsequently, in several papers including his dissertation, E. Stark$_{1,2,3}$ has made an extensive study of kernels with finite oscillation (see also C.J. Hoff$_1$) .

CHAPTER 5

SATURATION IN $C[a,b]$

5.1. Introduction. In this chapter, we will consider the saturation of positive operators in the space $C[a,b]$. The situation here differs considerably from the periodic case. For one thing, transform techniques which were so useful in the periodic case are no longer applicable. In fact, even the notion of convolution, which was the primary method of constructing operators in $C^*[-\pi,\pi]$ does not carry over as nicely to $C[a,b]$.

To make this point more clear, let's consider a sequence of non-negative continuous functions h_n and define the operators L_n on $C[a,b]$ by

$$L_n(f,x) = \int_a^b f(t)\, h_n(t-x)dt \; .$$

Consider the simplest function $f(t) = 1$ on $[a,b]$. Then

$$L_n(f,a) = \int_a^b h_n(t-a)dt = \int_0^{b-a} h_n(t)dt$$

and

$$L_n(f,b) = \int_a^b h_n(t-b)dt = \int_{a-b}^0 h_n(t)dt$$

If (L_n) is to be an approximating sequence then the functions h_n must concentrate their mass close to 0 as $n \to \infty$ and these masses must converge to 1. This forces either $\int_0^{b-a} h_n(t)dt$ or $\int_{a-b}^0 h_n(t)dt$ to converge to something $\leq \frac{1}{2}$ as $n \to \infty$ and thus

$$L_n(f,a) \not\to f(a) \quad \text{or} \quad L_n(f,b) \not\to f(b)$$

What we have pointed out is that such a sequence of operators cannot be used to approximate each function $f \in C[a,b]$ uniformly on $[a,b]$, because there is always difficulty at the end points of the interval. However, such a sequence is still useful for approximating on intervals $[c,d] \subsetneq [a,b]$ if $a < c < d < b$. We have seen at least one important example of this in the Landau operators introduced in 2.5.

Another common way of constructing operators in $C[a,b]$ is by an interpolation formula

$$L(f,x) = \sum_{k=1}^{m} f(x_k)\, h_k(x)$$

where $a \le x_1 < x_2 < \ldots < x_m \le b$ and for each k, h_k is a continuous and non-negative function on $[a,b]$. The Bernstein operators, Hermite-Fejér operators, and variation diminishing spline operators (see Chapter 2) are examples of this form.

5.2. Saturation in $C[a,b]$. The setup for the saturation problem in $C[a,b]$ is slightly different from that in C^*. Where as in C^*, we had that only the constant functions were approximated with precision "o" of the saturation order, we now wish to loosen this requirement by saying only that certain trivial functions are approximated with order "o" of the saturation order. This trivial class will depend on the sequence of operators and will have to be determined as part of the saturation problem. A second change is that we will now allow the saturation order to be a sequence of functions rather than just a sequence of real numbers. We need this in order to accommodate the possibility that the degree of approximation is effected by the position of the point x in the interval. For example, we have seen in Chapter 2 that approximation by Bernstein polynomials is better near the end points of the interval than it is for points in the interior.

Definition. A sequence of operators (L_n) *from* $C[a,b]$ *into* $C[c,d]$, $[c,d] \subseteq [a,b]$, *is said to be saturated on* $[c,d]$, *if there is a sequence of functions* $(\phi_n(x))$ *which tend to* 0 *uniformly on* $[c,d]$ *and are positive on* (c,d) *and a class of functions* $T(L_n)$ *such that*

5.2.1.

$$\lim_{n\to\infty} \|(\phi_n(x))^{-1}(f(x) - L_n(f,x))\| = 0$$

if and only if $f \in T(L_n)$ *and*

5.2.2. *there is a function* $f_0 \in C[a,b]$, $f_0 \notin T(L_n)$ *for which*

$$\|(\phi_n(x))^{-1}(f_0(x) - L_n(f_0,x))\| = O(1)$$

In the case that ϕ_n vanishes at either end point, the norm in (5.2.1) is to be understood as the supremum over the open interval (c,d). We let $S(L_n)$ denote the set of functions for which (5.2.2) holds.

Here too, we have dropped the $\underline{\lim}$ which appeared in the definition for the space C^*. Before, by using the $\underline{\lim}$ rather than just lim we were able to show that the saturation class did not depend on the saturation order. Now, because of the pointwise dependence of the saturation order, even if we incorporated the $\underline{\lim}$, we could not guarantee that $T(L_n)$ and $S(L_n)$ are independent of the saturation order.

This could have been avoided if we considered only saturation orders of the form $\phi_n(x) = \psi(x)\alpha_n$, with ψ a fixed function. In this case the returning to $\underline{\lim}$ definition we get a unique saturation class. We have given up knowing uniqueness of the saturation class for the advantage of obtaining somewhat more general saturation theorems. However, in all saturation theorems for classical operators, the saturation order can be taken as $\Psi(x)\alpha_n$ and in this sense the saturation class is unique.

It is still important to note that if (L_n) is saturated with order $(\phi_n(x))$ and if $(\Psi_n(x))$ is a sequence of functions for which there are constants $C_1, C_2 > 0$, with $C_1\Psi_n(x) \leq \phi_n(x) \leq C_2\Psi_n(x)$, for $x \in [c,d]$, then (L_n) is also saturated with order $(\Psi_n(x))$.

There is a distinction between the cases when $[c,d] = [a,b]$ and $a < c < d < b$. A saturation theorem in the latter case is called a local saturation theorem. In studying local saturation on an interval $[c,d]$, we stay away from the end points of the interval so that in this case we will take the functions $\phi_n(x)$ to be constant. Although this loses some generality, it suffices for all applications.

5.3. The Parabola Technique. We will begin our analysis with the problem of determining when a sequence (L_n) of positive operators is saturated, and if so, determining the saturation order. For the time being, we restrict our attention to techniques which apply to the case when $T(L_n)$ is the space of linear functions.

The following lemma of Bajsanski and Bojanic[1] is fundamental to most of the techniques we develop in this chapter.

Lemma 5.1. *Let* $f \in C[a,b]$ *with* $f(a) = f(b) = 0$ *and* $f(x_0) > 0$ *for some* $x_0 \in (a,b)$. *Then there is a quadratic polynomial* $Q(x) = \alpha x^2 + \beta x + \gamma$ *with* $\alpha < 0$, *such that*

$$Q(x) \geq f(x) \text{ on } [a,b] \tag{5.3.1}$$

and

$$Q(y) = f(y) \quad \text{for some} \quad y \in (a,b) \qquad (5.3.2.)$$

Remark. As we shall see in the proof, it is also possible to guarantee that $Q(a) > 0$ and $Q(b) > 0$.

Proof. Let $c = \frac{a+b}{2}$ and $M = 2\|f\|[a,b]$. Then for $\alpha < 0$ sufficiently close to 0 we have that the polynomial

$$Q_\alpha(x) = \alpha(x-c)^2 + M \geq f(x) \quad , \quad x \in [a,b] .$$

Also, choose $\alpha_o < 0$ sufficiently close to 0, so that we have

$$\alpha_o(\frac{b-a}{2})^2 + M > \alpha_o(x_o-c)^2 + M - f(x_o) .$$

This gives that

$$Q_{\alpha_o}(a) - f(a) > Q_{\alpha_o}(x_o) - f(x_o)$$

and

$$Q_{\alpha_o}(b) - f(b) > Q_{\alpha_o}(x_o) - f(x_o)$$

and therefore, the function $Q_{\alpha_o}(t) - f(t)$ assumes its minimum on $[a,b]$ somewhere in (a,b). Let this minimum be m. It is now easy to see that the polynomial $\alpha_o(x-c)^2 + M - m$ satisfies the conclusions of the lemma where y can be chosen as any point for which $Q_{\alpha_o}(y) - f(y) = m$.

A simple application of Lemma 5.1 gives the following "o" type saturation theorem.

Theorem 5.1. Let (L_n) *be a sequence of positive linear operators from* $C[a,b]$ *to* $C[a,b]$ *and define*

$$\mu_n(x) = \frac{1}{2} L_n((t-x)^2,x) .$$

If $\mu_n(x) > 0$, $x \in (a,b)$ *and for each* $x \in [a,b]$, $i = 0,1$

$$e_i(x) - L_n(e_i,x) = o_x(\mu_n(x)) \qquad (5.3.3)$$

then for $f \in C[a,b]$

$$f(x) - L_n(f,x) = o_x(\mu_n(x)) , \quad x \in (a,b) \qquad (5.3.4)$$

if and only if f *is linear on* $[a,b]$.

Remark. The subscript x in the notation o_x indicates that the o condition need only hold pointwise in x and not necessarily uniformly. Recall $e_i(x) = x^i$.

Proof. If f is linear, (5.3.3) guarantees that (5.3.4) holds. For the converse, let

$$g(x) = f(x) - f(a) - \frac{f(b) - f(a)}{b-a} (x-a) .$$

Then $g(a) = g(b) = 0$ and

$$g(x) - L_n(g,x) = o_x(\mu_n(x))$$

for each $x \in [a,b]$. This is because of (5.3.3).

Suppose $g(x_o) > 0$ for some $x_o \in (a,b)$. Using Lemma 5.1, we have that there is a point $y \in (a,b)$ and a quadratic $Q(x) = \alpha(x-y)^2 + \beta(x-y) + g(y)$, with $\alpha < 0$, for which $Q(x) \geq g(x)$ on $[a,b]$.

Now, we can estimate

$$L_n(g(x),y) - g(y) = L_n(g(x) - g(y),y) + L_n(g(y),y) - g(y)$$

$$= L_n(g(x) - g(y),y) + o_y(\mu_n(y))$$

where for the last equality we used (5.3.3).

Therefore,

$$L_n(g(x),y) - g(y) \leq L_n(Q(x) - Q(y),y) + o_y(\mu_n(y))$$

$$= \alpha\, L_n((x-y)^2,y) + \beta\, L_n(x-y,y) + o_y(\mu_n(y)) = 2\alpha\, \mu_n(y) + o_y(\mu_n(y))$$

where again we used (5.3.3) and the definition of $\mu_n(y)$.

Since $\alpha < 0$, we see that

$$L_n(g(x),y) - g(y) \neq o_y(\mu_n(y))$$

and this is a contradiction. Thus, we must have $g(x) \leq 0$ for all $x \in [a,b]$. But, if $g(x_o) < 0$, for some $x_o \in (a,b)$, then considering the function $-g$ instead of g , we again arrive at a contradiction. This shows that $g(x) = 0$ on $[a,b]$ and thus f is linear.

Theorem 5.1 gives a result somewhat stronger than the "o" part of the definition of saturation. The reason is that we have required that $f(x) - L_n(f,x) = o_x(\mu_n(x))$ only holds pointwise and still were able to conclude that f is linear. To obtain the saturation condition (5.2.2) we need only require that (5.3.3) hold uniformly in x .

Theorem 5.2. *Let* (L_n) *be a sequence of positive operators from* $C[a,b]$ *to* $C[a,b]$ *and define* $\mu_n(x)$ *as in Theorem 5.1. If for* $i = 0,1$

$$\left\| \frac{e_i(x) - L_n(e_i,x)}{\mu_n(x)} \right\| = o(1) \tag{5.3.5}$$

Then (L_n) *is saturated with order* $(\mu_n(x))$ *and trivial class* $T(L_n) = \{ \ell : \ell$ *is linear on* $[a,b]\}$.

Proof. The "o" condition (5.2.1) follows from Theorem 5.1. Using (5.3.5) , it is easy to show that the function $f_o(t) = t^2$ satisfies the "O" condition (5.2.2).

As an example, we can determine a saturation order for Bernstein polynomials. From 2.7, we see that $B_n(e_i) = e_i$, $i = 0,1$.

Also,

$$\mu_n(x) = \frac{1}{2} B_n((t-x)^2,x) = \frac{x(1-x)}{2n} \tag{5.3.6}$$

Therefore (B_n) is saturated with order $\left(\frac{x(1-x)}{2n}\right)$ and trivial class $T(B_n) = \{ \ell : \ell$ is linear on $[0,1]\}$.

5.4. A Local "o" Saturation Theorem. In order to put Theorem 5.1 in the form of a local saturation theorem, we need some additional assumptions on (L_n).

Theorem 5.3. Let (L_n) *be a sequence of positive linear operators from* $C[a,b]$ *into* $C[c,d]$ *with* $a < c < d < b$ *and*

$$0 < \mu_n = \frac{1}{2} \| L_n((t-x)^2,x) \|$$

Suppose that (L_n) satisfies the following three conditions:

(5.4.1) $$\| e_i - L_n(e_i) \| = o(\mu_n) \qquad i = 0,1 \quad .$$

(5.4.2) There are positive constants C_1 and C_2 such that

$$C_1\mu_n \leq L_n((t-x)^2,x) \leq C_2 \mu_n \text{ , for each } x \in [c,d] \text{ , } n = 1,2,\ldots$$

(5.4.3) If $f \in C[a,b]$ vanishes on $[c',d']$ with $c' < c$ and $d < d'$ then

$$\| f - L_n(f) \| = o(\mu_n) \ .$$

Then, (L_n) is saturated on $[c,d]$ with order (μ_n) and trivial class $T(L_n) = \{ \ell : \ell$ is linear on $[c,d]\}$.

Remark. As usual the $\| \cdot \|$ is taken in the image space and is thus the supremum over $[c,d]$. The condition (5.4.2) is needed, since, in the parabola technique, we have no control over where the point y of Lemma 5.1 is located. The condition (5.4.3) is used to neglect the contribution on $[a,b] \setminus [c,d]$. If

$$\| L_n((t-x)^4,x) \| [c,d] = o(\mu_n) \ ,$$

then (5.4.3) will hold. (see Lemma 5.2).

Proof. Suppose $f \in C[a,b]$ and

$$\| f - L_n(f) \| = o(\mu_n) .$$

Let $g = f - \ell$, where ℓ is the linear function which interpolates f at c and d . Then $\| g - L_n(g) \| = o(\mu_n)$ follows from (5.4.1). We wish to show that $g(x) = 0$ for $x \in [c,d]$. Suppose $g(x_o) > 0$ for some $x_o \in (c,d)$. Using Lemma 5.1, we see that there is a point $y \in (c,d)$ and a quadratric $Q(x) = \alpha(x-y)^2 + \beta(x-y) + g(y)$ such that

$$Q(x) \geq g(x) \quad \text{on} \quad [c,d]$$

Also, as remarked after Lemma 5.1, this quadratic can be chosen so that $Q(c) > 0$ and $Q(d) > 0$.

Let $c' = \max \{ x : x < c \text{ and } Q(x) = g(x)\}$. Where there are no points for which $Q(x) = g(x)$ in $[a,c]$, we let $c' = a$, then $c' < c$. Similarly, we let $d' = \min \{ x : x > d \text{ and } g(x) = Q(x)\}$. We define $h \in C[a,b]$ by $h(x) = 0$ on $[c',d']$ and $h(x) = g(x) - Q(x)$ on $[a,b] - [c',d']$. Thus,

$$g(x) \leq Q(x) + h(x) \quad \text{for} \quad x \in [a,b]$$

This gives by virtue of (5.4.1) that

$$L_n(g(x),y) - g(y) = L_n(g(x) - g(y), y) + L_n(g(y),y) - g(y)$$

$$\leq L_n(Q(x) + h(x) - Q(y) - h(y), y) + o(\mu_n)$$

$$= \alpha\, L_n((x-y)^2,y) + L_n(h(x) - h(y),y) + o(\mu_n)$$

But, h vanishes on $[c',d']$ so that from (5.4.3) we find

$$L_n(h(x) - h(y),y) = o(\mu_n)$$

Therefore,

$$L_n(g(x),y) - g(y) \leq \alpha\, L_n((x-y)^2,y) + o(\mu_n)$$

Since $\alpha < 0$, we have $L_n((x-y)^2,y) \neq o(\mu_n)$, because of (5.4.2). This gives the desired contradiction and thus $g(x) \leq 0$ on $[c,d]$. To see that $g(x) \geq 0$ on $[c,d]$, we merely replace g by $-g$ in the above argument.

The function $f_o(t) = t^2$ satisfies $\| f_o - L_n(f_o) \| = O(\mu_n)$. This shows that L_n is locally saturated on $[c,d]$ with order (μ_n) and trivial class $T(L_n) = \{ \ell : \ell$ is linear on $[c,d]\}$.

For convolution type operators, the conditions of Theorem 5.3 can be replaced by simple conditions on the moments of the kernels.

Corollary 5.1. *Let (L_n) be a sequence of positive operators defined on $C[-\frac{1}{2},\frac{1}{2}]$ by the convolution formula*

$$L_n(f,x) = \int_{-\frac{1}{2}}^{\frac{1}{2}} f(t)\, d\alpha_n(t-x)$$

where for each $n, d\alpha_n$ is a non-negative, even, Borel measure on $[-1,1]$ with

$$\int_{-1}^{1} d\alpha_n(t) = 1 . \tag{5.4.4}$$

Let

$$\mu_n = \int_{-1}^{1} t^2\, d\alpha_n(t) . \tag{5.4.5}$$

If

$$\int_{-1}^{1} t^4 \, d\alpha_n(t) = o(\mu_n) \tag{5.4.6}$$

and $0 < \delta < \frac{1}{2}$, *then* (L_n) *is locally saturated on* $[-\delta,\delta]$ *with order* (μ_n) *and trivial class* $T(L_n) = \{ \ell : \ell$ *is linear on* $[-\delta,\delta]\}$.

Proof. We have for $e_o(x) = 1$,

$$\| e_o - L_n(e_o) \| = \| 1 - \int_{-\frac{1}{2}}^{\frac{1}{2}} d\alpha_n(t-x) \|$$

$$= \| \int_{-1}^{1} d\alpha_n(t) - \int_{-\frac{1}{2}-x}^{\frac{1}{2}-x} d\alpha_n(t) \|$$

$$\leq \| [\int_{-1}^{-\frac{1}{2}-x} + \int_{\frac{1}{2}-x}^{1}] d\alpha_n(t) \| \leq 2 \int_{\frac{1}{2}-\delta}^{1} d\alpha_n(t)$$

Since $(\frac{1}{2} - \delta)^{-4} t^4 \geq 1$ for $t \geq (\frac{1}{2} - \delta)$ we have

$$\| e_o - L_n(e_o) \| \leq 2\int_{\frac{1}{2}-\delta}^{1} d\alpha_n(t) \leq (\frac{1}{2} - \delta)^{-4} \int_{-1}^{1} t^4 \, d\alpha_n(t) = o(\mu_n)$$

where, we have used (5.4.6). Similarly, we can estimate $\| e_1 - L_n(e_1) \|$, using this same idea. Now, we use the fact that $d\alpha_n$ is an even measure.

So that,

$$\| e_1 - L_n(e_1) \| = \| x \int_{-1}^{1} d\alpha_n(t) - \int_{-\frac{1}{2}-x}^{\frac{1}{2}-x} (x+t) d\alpha_n(t) \|$$

$$\leq 2 \int_{\frac{1}{2}-\delta}^{1} d\alpha_n(t) + 2 \int_{\frac{1}{2}-\delta}^{1} |t| \, d\alpha_n(t)$$

$$\leq 4 \int_{\frac{1}{2}-\delta}^{1} d\alpha_n(t) = o(\mu_n)$$

This shows that (5.4.1) is valid.

The estimate

$$C_1\mu_n \le L_n((t-x)^2,x) \le C_2\mu_n \qquad \text{for } x \in [-\delta,\delta]$$

follows easily from the fact that $\int_{\frac{1}{2}-\delta}^{1} d\alpha_n(t) = o(\mu_n)$.

Finally, to show (5.4.3) suppose f vanishes on $[-\delta',\delta']$ with $\frac{1}{2} \ge \delta' > \delta$ then

$$\| f - L_n(f) \| = \left\| \left[\int_{-\frac{1}{2}}^{-\delta'} + \int_{\delta'}^{\frac{1}{2}} \right] f(t)\, d\alpha_n(t-x) \right\|$$

$$\le 2M \int_{\delta'-\delta}^{1} d\alpha_n(t) \le \frac{2M}{(\delta'-\delta)^4} \int_{-1}^{1} t^4\, d\alpha_n(t) = o(\mu_n)$$

where $M = \|f\|\ [-\frac{1}{2},\frac{1}{2}]$.

As a consequence of Corllary 5.1 and the estimates of 2.5.4, we see that for each $0 < \delta < \frac{1}{2}$ the sequence of Landau operators (ℓ_n) is locally saturated on $[-\delta,\delta]$ with order (n^{-1}) and trivial class $T(\ell_n) = \{\ell : \ell$ is linear on $[-\delta,\delta]\}$.

5.5. The Saturation Class. We can also use the parabola technique to obtain a characterization of the saturation class. This approach had its origin in the work of V. Amelkovic[1] and subsequently G. Muhlbach[1]. In order to use the parabola technique to determine the saturation class, we of course need to assume that (L_n) satisfies the hypothesis of the "o" theorem (Theorem 5.2). Also, we will need the following additional property.

If f vanishes on an open set $\theta \subseteq [a,b]$ and $y \in \theta$ then

$$f(y) - L_n(f,y) = o_y(\mu_n(y))$$

This condition essentially says that the contribution to $L_n(f,y)$ outside of a neighborhood of y is $o_y(\mu_n(y))$. It is the analogue of (3.9.4) which formed part of the hypothesis of Tureckiĭ's Theorem, pg. 69.

It, too, has many equivalent formulations. We summarize the important ones in the following lemma.

Lemma 5.2. *Let* (L_n) *be a sequence of positive operators mapping* $C[a,b]$ *into* $C[c,d]$. *Furthermore, let* $y \in [c,d]$ *and define* $\mu_n(y)$ *as in Theorem* 5.1. *If* (5.3.3) *holds at* y *then the following are equivalent.*

(5.5.1) *If* f *is any function in* $C[a,b]$ *which vanishes on some open subset* θ *of* $[a,b]$ *and* $y \in \theta$ *then*

$$L_n(f,y) = o(\mu_n(y))$$

(5.5.2)

$$L_n((t-y)^4,y) = o(\mu_n(y))$$

(5.5.3) *If* $f \in C[a,b]$ *is twice continuously differentiable in some neighborhood of* y *then*

$$\lim_{n\to\infty} \frac{L_n(f,y) - f(y)}{\mu_n(y)} = f''(y)$$

Remark. Note that (5.5.3) implies (5.3.3) but we need to assume (5.3.3) to prove the other equivalences.

Proof. For each $y \in [c,d]$, (5.5.3) clearly implies (5.5.2) since the function $f(t) = (t-y)^4$ is twice continuously differentiable and $f''(y) = 0$. To see that (5.5.2) implies (5.5.1) suppose f vanishes on $\theta \subseteq [a,b]$ and $y \in \theta$. For M suitably large, $M(t-y)^4 \geq |f(t)|$ on $[a,b]$. Thus

$$|L_n(f(t),y)| \leq M\, L_n((t-y)^4,y) = o(\mu_n(y)) \quad .$$

Finally, we wish to show that (5.5.1) implies (5.5.3) . To see this, suppose f is twice continuously differentiable in a neighborhood of y . Then when $t \in [a,b]$, we have

$$f(t) = f(y) + f'(y)\,(t-y) + \frac{1}{2}\, f''(y)\,(t-y)^2 + \epsilon(\,t-y\,)\,(t-y)^2$$

where $\epsilon(t)$ is a continuous function which vanishes at 0 .
Therefore,

$$L_n(f(t),y) - f\,(y) = L_n(f(y),y) - f(y) + f'(y)\, L_n((t-y),y)$$

$$+ f''\,(y)\mu_n(y) + L_n(\varepsilon(\,t-y\,)\,(t-y)^2,y) = f''(y)\mu_n(y)$$

$$+ L_n(\varepsilon(\,t-y\,)\,(t-y)^2,y) + o(\mu_n(y)) \qquad (5.5.4)$$

where we have used (5.3.3). Given $\varepsilon_0 > 0$ choose an open set θ containing y, so $|\varepsilon(\,t-y\,)| < \varepsilon_0$ when $t \in \theta$. Then ,

$$|\varepsilon(\,t-y\,)\,(t-y)^2| \leq \varepsilon_0(t-y)^2 + \max\,[0,\varepsilon(\,t-y\,) - \varepsilon_0]\,(t-y)^2$$

$$\leq \varepsilon_0(t-y)^2 + g(t)$$

where g vanishes on θ . Thus,

$$|L_n(\varepsilon(\ t-y\ )\ (t-y)^2,y)| \leq \varepsilon_o L_n((t-y)^2,y) + L_n(g(t),y)$$

$$= 2\varepsilon_o \mu_n(y) + L_n(g(t),y) \leq 3\varepsilon_o \mu_n(y)$$

for n sufficiently large, by virtue of (5.5.1). Using this in (5.5.4), and dividing by $\mu_n(y)$ gives for n sufficiently large

$$|\mu_n^{-1}(y)\ (L_n(f,y) - f(y)) - f''(y)| \leq 3\varepsilon_o + o(1)\ .$$

which proves (5.5.3), since ε_o was arbitrary.

The condition (5.5.3) is commonly known as the Voronovskaya condition. E. Voronovskaya$_1$ showed this for the Bernstein operators. The equivalence of (5.5.3) and (5.5.2) can be found in R. Mamedov$_2$.

The following theorem, which was given by G. Muhlbach$_1$ determines the saturation class for a sequence of operators which satisifes one of the conditions in Lemma 5.2, for each $y \in [a,b]$.

<u>Theorem 5.4.</u> <u>Let</u> (L_n) <u>be a sequence of positive operators from</u> $C[a,b]$ <u>to</u> $C[a,b]$ <u>and let</u>

$$\mu_n(x) = \frac{1}{2} L_n((t-x)^2,x)\ .$$

<u>be positive on</u> (a,b) . Suppose that for $i = 0,1$ we have

$$e_i(x) - L_n(e_i,x) = o_x(\mu_n(x)) \qquad x \in [a,b] \qquad (5.5.5)$$

If (L_n) *satisfies any of the equivalent conditions in Lemma* 5.2, *for each point* $y \in (a,b)$, *then*

$$|f(x) - L_n(f,x)| \leq M\mu_n(x) + o_x(\mu_n(x)) \qquad x \in (a,b) \qquad (5.5.6)$$

if and only if $f' \in \text{Lip}\,(1,M)$

Remarks. If in addition (5.3.5) holds, then (L_n) is saturated with order $(\mu_n(x))$ by virtue of Theorem 5.2. If the equivalent conditions of Lemma 5.2 hold uniformly in x then Theorem 5.4 then shows that $S(L_n) = \{f : f' \in \text{Lip}\ 1\}$, since (5.5.6) will then hold uniformly in x . Of course Theorem 5.4 gives a sharp characterization of $S(L_n)$ since the analysis includes the dependence on the constant M . When $L_n(e_i) = e_i$, $i = 0,1$ then the o terms in (5.5.5) and (5.5.6) can be dropped.

Proof. If $f' \in \text{Lip}\,(1,M)$ then the estimate (5.5.6) follows from Theorem 2.5. Now suppose $f \in C[a,b]$ and for each $x \in (a,b)$

$$|f(x) - L_n(f,x)| \leq M\mu_n(x) + o_x(\mu_n(x))$$

We suppose further that f' is not in $\text{Lip}(1,M)$ and work for a contradiction. If $f' \notin \text{Lip}(1,M)$ then for some $x_o \in (a,b)$ and $0 < \delta < \min(|a-x_o|, |b-x_o|)$, we have

$$\left|\frac{f(x_o+\delta) + f(x_o-\delta) - 2f(x_o)}{\delta^2}\right| > M$$

This follows from the characterization (1.3.6) . We will suppose that

$$M'\delta^2 = f(x_o+\delta) + f(x_o-\delta) - 2f(x_o) < -M\,\delta^2$$

The other case is handled similarly (or can be gotten by replacing f by $-f$) Consider the quadratic function $Q(t) = -\frac{1}{2}\,\alpha\,(t-x_o)^2 + \ell(t) + C$ where $\alpha = \frac{M+M'}{2}$, ℓ is the linear function which interpolates f at $x_o - \delta$ and $x_o + \delta$, and C is chosen large enough that $Q(t) \geq f(t)$ on $[x_o-\delta,\ x_o+\delta]$. Then at $x_o - \delta$ we have

$$Q(x_o-\delta) - f(x_o-\delta) = -\frac{1}{2}\alpha\,\delta^2 + \ell(x_o-\delta) + C - f(x_o-\delta) = -\frac{1}{2}\alpha\delta^2 + C$$

Similarly

$$Q(x_o+\delta) - f(x_o+\delta) = -\frac{1}{2}\alpha\,\delta^2 + C \quad .$$

While at the point x_o we have

$$Q(x_o) - f(x_o) = C + \ell(x_o) - f(x_o)$$

$$= C + f(x_o-\delta) + \frac{[f(x_o+\delta) - f(x_o-\delta)](x_o-x_o+\delta)}{2\delta} - f(x_o)$$

$$= C + \frac{f(x_o-\delta) + f(x_o+\delta) - 2f(x_o)}{2}$$

$$= C - \frac{M'}{2}\,\delta^2 < C - \frac{1}{2}\alpha\,\delta^2$$

Thus

$$Q(x_o) - f(x_o) < Q(x_o-\delta) - f(x_o-\delta)$$

and

$$Q(x_o) - f(x_o) < Q(x_o+\delta) - f(x_o+\delta)$$

This shows that

$$m = \inf\{Q(t) - f(t) : t \in [x_o-\delta, x_o+\delta]\}$$

is assumed at a point y in the interior of $[x_o - \delta , x_o + \delta]$. We let $Q^*(t) = Q(t) - m$, then Q^* has the following properties

$$Q^*(t) \geq f(t) \quad \text{for} \quad t \in [x_o - \delta, x_o + \delta] \tag{5.5.7}$$

$$Q^*(y) = f(y) \tag{5.5.8}$$

Let

$$a' = \max \{x : x \leq x_o - \delta, Q^*(x) = f(x)\} \text{ and}$$

$$b' = \min \{x : x \geq x_o + \delta, Q^*(x) = f(x)\} .$$

then

$$a \leq a' < y < b' \leq b \quad \text{and} \quad Q^*(t) \geq f(t) \quad \text{on} \quad [a',b'] .$$

We let $g(t) = \begin{cases} 0, & t \in [a',b'] \\ f(t) - Q^*(t), & t \notin [a',b'] \end{cases}$

so that $f(t) \leq Q^*(t) + g(t)$, $t \in [a,b]$.

The function g is in $C[a,b]$ and by virtue of (5.5.1) we have

$$L_n(g,y) - g(y) = o_y(\mu_n(y)) \ .$$

Therefore, using this remark together with (5.5.5) shows that

$$L_n(f,y) - f(y) = L_n(f(t) - f(y),y) + o_y(\mu_n(y))$$

$$\leq L_n(Q^*(t) + g(t) - Q^*(y) - g(y),y) + o_y(\mu_n(y))$$

$$= L_n(Q^*(t) - Q^*(y),y) + o_y(\mu_n(y))$$

$$= - \frac{\alpha}{2} L_n((t-y)^2,y) + o_y(\mu_n(y)) < - M \mu_n(y) + o_y(\mu_n(y))$$

This means that (5.5.6) can't hold at y, and we have a contradiction which proves the thoerem.

Corollary 5.2 . The Bernstein operators are saturated with order $(\frac{x(1-x)}{2n})$ and trivial class $T(B_n) = \{\ell : \ell$ is linear on $[0,1]\}$. If $f \in C[0,1]$ then

$$|f(x) - B_n(f,x)| \leq M \frac{x(1-x)}{2n} \qquad (5.5.9)$$

if and only if $f' \in Lip(1,M)$

Remark. In (5.5.9) we are able to drop the term $o_x(\mu_n(x))$ since for the Bernstein operators $B_n(e_i) = e_i$, $e_i = 0,1$, $n \in N$.

Proof. From formulas (2.7.2) of Chapter 2 , it easily follows that

$$B_n((t-x)^4,x) = 3n^{-2}x^2(1-x)^2 + n^{-3}x(1-x)[1-6x+6x^2] = o(x(1-x)n^{-1})$$

with the "o" uniform in $0 \leq x \leq 1$. Thus, (5.5.2) of Lemma 5.2 holds. The corollary then follows from Theorem 5.4.

This result is due to G.G. Lorentz[1] who improved upon a result of K. De Leeuw[1] The theorem of DeLeeuw did not take into consideration the behavior near the end points.

We can also determine the saturation for variation diminishing splines provided the degree of the splines does not go to infinity too fast in comparison with the mesh length.

Corollary 5.3. *Let* (s_{Δ_n,m_n}) *be a sequence of variation diminishing spline operators mapping* $C[0,1]$ *into* $C[0,1]$. *If* $m_n \| \Delta_n \| \to 0$ $(n \to \infty)$, *then* (s_{Δ_n,m_n}) *is saturated with order*

$$\mu_n(x) = \frac{1}{2} s_{\Delta_n,m_n}((t-x)^2,x)$$

and trivial class $T(s_{\Delta_n,m_n}) = \{\ell : \ell$ *is linear on* $[0,1]\}$. *If* $f \in C[0,1]$ *then*

$$|f(x) - s_{\Delta_n,m_n}(f,x)| \leq M\, \mu_n(x) \qquad 0 \leq x \leq 1$$

if and only if $f' \in Lip(1,M)$.

Proof. This will follow from Theorem 5.4, if we can show that

$$s_{\Delta_n,m_n}((t-x)^4,x) = o(\mu_n(x)) \qquad (n \to \infty) \tag{5.5.10}$$

uniformly in $0 \leq x \leq 1$.

We already know that s_{Δ_n,m_n} preserves linear functions .

The normalized B spline N_i has support on $[x_i , x_{i+m_n}]$. Thus if $x \notin [x_i , x_{i+m_n}]$ then $N_i(x) = 0$. Whenever i is an index with $x \in [x_i , x_{i+m_n}]$ then

$$|\xi_i - x| = \left| \frac{(x_{i+1}-x)+\ldots+(x_{i+m_n-1}-x)}{m_n-1} \right| \leq m_n \| \Delta_n \| \tag{5.5.11}$$

Thus,

$$(\xi_i - x)^4 \leq m_n^2 \| \Delta_n \|^2 (\xi_i - x)^2$$

This shows that

$$s_{\Delta_n,m_n}((t-x)^4,x) = \sum (\xi_i - x)^4 N_i(x)$$

$$\leq m_n^2 \| \Delta_n \|^2 \sum (\xi_i - x)^2 N_i(x) = m_n^2 \| \Delta_n \|^2 \cdot (2\mu_n(x)) = o(\mu_n(x))$$

uniform in $[0,1]$. This is (5.5.10) as desired. Note, we needed only to sum over those indices where $N_i(x) \neq 0$ and therefore (5.5.11) can be used.

5.6. The Saturation Class for Local Saturation. A "O" local saturation theorem can be proved using the same hypothesis as for the "o" theorem, except that, we will also need that one of the conditions of Lemma 5.2. holds.

Theorem 5.5. Let L_n be a sequence of positive operators mapping $C[a,b]$ into $C[c,d]$ which satisfies the hypothesis of Theorem 5.3 and also satisfies the equivalent conditions of Lemma 5.2 uniformly on $[c,d]$. Then $S(L_n) =$ $= \{f : f' \in \text{Lip } 1 \text{ on } [c,d]\}$.

Remark. We need not assume (5.4.3) since it is implied by (5.5.3) .

Proof. If $f' \in \mathrm{Lip}\ 1$, then $f \in S(L_n)$. This follows from Theorem 2.3. Now suppose $f \in S(L_n)$ so that

$$\| f - L_n(f) \| \leq M \mu_n \qquad n = 1,2,\ldots \tag{5.6.1}$$

for some $M > 0$. We will show that $f' \in \mathrm{Lip}(1, MC_1^{-1})$ where C_1 is the constant in 5.4.2. To see this, we argue exactly as in Theorem 5.4 . Suppose that for some $x_o \in [c,d]$

$$\left| \frac{f(x_o+\delta) + f(x_o-\delta) - 2f(x_o)}{\delta^2} \right| > MC_1^{-1}$$

and say

$$\frac{f(x_o+\delta) + f(x_o-\delta) - 2f(x_o)}{\delta^2} < -M\,C_1^{-1}$$

Arguing as in the proof of Theorem 5.4, we can find a point $y \in (c,d)$ an interval $[c',d'] \subseteq [c,d]$ with $c' < y < d'$, a quadratic $Q^*(x) = -\frac{1}{2}\alpha(x-y)^2 + \beta(x-y) + f(y)$ with

$$-\alpha < -MC_1^{-1} \tag{5.6.2}$$

and a function g which vanishes on $[c',d']$ such that

$$f(t) \leq Q^*(t) + g(t) , \qquad t \in [a,b] .$$

Then

$$L_n(f,y) - f(y) \leq -\frac{1}{2}\alpha\, L_n((t-y)^2,y) + o(\mu_n) < -M\,\mu_n + o(\mu_n)$$

Where in the first inequality we have used (5.4.1) and (5.5.1). The second inequality follows from (5.6.2) and (5.4.2) . This contradicts (5.6.1), as desired.

Corollary 5.4. Let (L_n) satisfy the hypothesis of Corollary 5.1. For each interval $[-\delta,\delta]$ with $\delta < \frac{1}{2}$, the sequence of operators (L_n) is locally saturated on $[-\delta,\delta]$ with order (μ_n) trivial class $T(L_n) = \{\ell : \ell$ is linear on $[-\delta,\delta]\}$, and saturation class $S(L_n) = \{f : f' \in \text{Lip } 1$ on $[-\delta,\delta]\}$.

Proof. Everything has been verified in the proof of Corollary 5.1 except that one of the conditions of Lemma 5.2. holds. But, (5.5.2) follows easily from (5.4.6) .

In particular, Corollary 5.4 determines the saturation for the Landau operators.

Corollary 5.5. For each $\delta < \frac{1}{2}$, the sequence of Landau operators (ℓ_n) is locally saturated on $[-\delta,\delta]$ with order (n^{-1}), trivial class $T(\ell_n) =$ $= \{\ell : \ell$ is linear on $[-\delta,\delta]\}$ and saturation class $S(\ell_n) = \{f : f'$ is in Lip 1 on $[-\delta,\delta]\}$

Proof. The estimates of 2.5.4 show that (ℓ_n) satisfies the hypothesis of Cor. 5.1.

Corollary 5.4. can also be used to obtain local saturation theorems in the periodic case. For example, the following corollary applies to operators which satisfy the hypothesis of Tureckiĭ's Theorem (Theorem 3.6).

Corollary 5.6. Let (L_n) be a sequence of convolution operators on C^* which satisfy the hypothesis of Theorem 3.6. If $\delta < \pi$, then (L_n) is locally saturated on $[-\delta,\delta]$ with order $(1-\rho_{1,n})$, trivial class $T(L_n)=\{\ell:\ell$ is linear on

$[-\delta,\delta]\}$ <u>and saturation class</u> $S(L_n) = \{f : f' \in \text{Lip } 1$ <u>on</u> $[-\delta,\delta]\}$.

<u>Proof.</u> Each L_n can be naturally extended to continuous non-periodic functions on $[-\pi,\pi]$ by means of the formula

$$L_n(f,x) = \int_{-\pi}^{\pi} f(t)\, d\mu_n(t-x)$$

We need to verify the hypothesis of Corollary 5.4, which in turn are the hypothesis of Corollary 5.1. The crucial thing here is that we know

$$\int_{\complement\theta} d\mu_n(t) = o(1-\rho_{1,n})$$

for each open neighborhood θ of 0 . From this it follows that (5.4.6) holds where

$$\int_{-\pi}^{\pi} t^2\, d\mu_n(t) \sim 1 - \rho_{1,n}$$

<u>5.7. Pointwise Saturation.</u> Theorem 5.1 as we have observed is actually a pointwise theorem, since, if

$$f(x) - L_n(f,x) = o_x(\mu_n(x)) \tag{5.7.1}$$

pointwise on $[a,b]$ then f is linear. It is then natural to attempt to characterize the functions f which satisfy a corresponding pointwise " O " condition, namely,

$$f(x) - L_n(f,x) = O_x(\mu_n(x)) , \qquad x \in [a\ b] \tag{5.7.2}$$

We can restate this in a slightly more general form: characterize the functions f for which

$$\underline{\lim_{n\to\infty}} (\mu_n(x))^{-1}[L_n(f,x) - f(x)] \le g(x) \le \overline{\lim_{n\to\infty}} (\mu_n(x))^{-1}[L_n(f,x) - f(x)] \qquad (5.7.3)$$

$$x \in (a,b)$$

where g is a finite valued function on $[a,b]$. Here, of course, if (5.7.2) holds we would take $g(x)$ to be either the right hand side or the left hand side of (5.7.3) in order to recover (5.7.2). See also the remarks on page 153.

Our approach to this problem will be that of H. Berens[4]. As we shall see, the technique used is essentially the same as the parabola technique given in Sections 5.5 and 5.6. However, looked at in a different way. This new viewpoint is much more amenable to the generalizations which are given in the next section.

For orientation purposes, let's begin with a simple sequence of operators. For $t > 0$, let $L_t(f,x) = \frac{1}{2}(f(x+t) + f(x-t))$. We have determined the saturation of the sequence $(L_{n^{-1}})_{n=1}^{\infty}$ previously in Chapter 3 by using the transform technique. The sequence (L_t) is saturated with order $\frac{1}{2}t^2$ and saturation class $\{f : f' \in \text{Lip } 1\}$. ($(\frac{1}{2}t^2)$ is chosen as the saturation order so that (5.5.3) will hold for L_t). But now, in a pointwise form, we would ask to characterize those functions f for which

$$\underline{\lim_{t\to o}}\, 2t^{-2}[L_t(f,x) - f(x)] \le g(x) \le \overline{\lim_{t\to o}}\, 2t^{-2}[L_t(f,x) - f(x)] \qquad (5.7.4)$$

$$x \in [a,b]$$

The left hand side and right hand side of (5.7.4) are lower $\underline{D}_2(f,x)$ and upper $\overline{D}_2(f,x)$ second symmetric derivatives of f. So that, (5.7.4) can be restated as

$$\underline{D}_2(f,x) \leq g(x) \leq \overline{D}_2(f,x) \ , \quad x \in (a,b) \qquad (5.7.5)$$

The characterization desired is given in the following classical lemma of de la Valee Poussin .

Lemma 5.3. *Let* $f \in C[a,b]$ *and* g *be a finitely valued function in* $L_1[a,b]$. *If* f *satisfies* (5.7.5), *then*

$$f(x) = A + Bx + \int_a^x \int_a^t g(u)du\,dt \qquad a \leq x \leq b \ . \qquad (5.7.6)$$

where A *and* B *are constants.*

Remark. The characterization (5.7.6) contains the saturation of (L_t) (given in Theorem 5.4) since when (5.5.5) holds, then we can take $|g(x)| \leq M$ on $[a,b]$.

Proof. The most important observation to make is that

$$\text{if} \quad \overline{D}_2(f,x) \geq 0 \quad \text{on} \quad [a,b] \ , \quad \text{then} \quad f \quad \text{is convex.} \qquad (5.7.7)$$

Of course, this is a well known classical result, but it is here that the parabola technique enters the picture. Let's see how the parabola technique can give us (5.7.7). Suppose for some $x_o, y_o \in [a,b]$ and $x_o < \xi < y_o$, we had the point $(\xi, f(\xi))$ lies above the line $(x, \ell(x))$ connecting $(x_o, f(x_o))$ to $(y_o, f(y_o))$. Then the function $f - \ell$ would vanish at x_o and y_o and also we would have $f(\xi) - \ell(\xi) > 0$. By Lemma 5.1, there is a quadratic polynomial $Q(x) = \alpha x^2 + \beta x + \gamma$

with $\alpha < 0$ such that $Q(x) \geq f(x) - \ell(x)$ on $[x_0,y_0]$ and $Q(\eta) = f(\eta) - \ell(\eta)$ for some point $\eta \in (x_0,y_0)$. But then, for t sufficiently small

$$L_t(Q,\eta) - Q(\eta) \geq L_t(f,\eta) - f(\eta) \quad .$$

Dividing by $\frac{1}{2}t^2$ and taking a $\overline{\lim_{t\to 0}}$, we see that

$$0 > 2\alpha = \overline{D}_2(Q,\eta) \geq \overline{D}_2(f,\eta)$$

This gives the desired contradiction and shows (5.7.7).

Now to prove the lemma, we follow the argument in A.Zygmund$_1$ [Vol. I, p. 327] . First, we introduce some notation. Let $H(x) = f(x) - \int_a^x \int_a^t g(\mu)d\mu\, dt$ and for $n \in N$ let M_n be a major function for g , such that $|M_n(x) - \int_a^x g(\mu)\, d\mu| \leq n^{-1}$, $x \in [a,b]$. To say M_n is a major function means the Dini numbers for M_n at x are larger than or equal to $g(x)$. The existence of such functions is well known from the theory of Lebesgue integration (see I. Pesin$_1$, Theorem 9.1, page 160). In particular, it follows that if $M_n^*(x) = \int_a^x M_n(t)dt$ then

$$\underline{D}_2(M_n^*,x) \geq g(x)$$

This can be shown directly but it is also contained in Lemma 5.5. Similarly, let m_n be a minor function for g and $m_n^*(x) = \int_a^x m_n(t)dt$ where we require that for $n \in N$

$$|\int_a^x g(\mu)d\mu - m_n(x)| < \frac{1}{n}, \quad x \in [a,b] .$$

We then have

$$\overline{D}_2(f,x) \geq g(x) \geq \overline{D}_2(m_n^*,x)$$

So that

$$\overline{D}_2(f-m_n^*,x) \geq \overline{D}_2(f,x) - \overline{D}_2(m_n^*,x) \geq 0$$

This shows that for each n, $f - m_n^*$ is convex. Taking a limit, we see that H is convex. In a similar way, we find $\underline{D}^2(f - M_n^*,x) \leq 0$ on $[a,b]$, so that H is concave. Since H is both convex and concave on $[a,b]$, H is linear on $[a,b]$ and the theorem is proved.

Now, we want to use the same kind of argument to obtain a general pointwise "0" saturation theorem for operators that satisfy an asymptotic condition (5.5.3). In place of (5.5.3) we could require (5.5.1) or (5.5.2) together with (5.3.3) on $[a,b]$.

<u>Lemma 5.4.</u> <u>Let</u> (L_n) <u>be a sequence of positive linear operators mapping</u> $C[a,b]$ <u>into</u> $C[a,b]$ <u>and satisfying</u> (5.5.3) <u>on</u> $[a,b]$. <u>If</u> $f \in C[a,b]$ <u>and</u>

$$\overline{\lim_{n\to\infty}}\, (\mu_n(x))^{-1}\, [L_n(f,x) - f(x)] \geq 0\,, \qquad x \in (a,b)$$

<u>then</u> f <u>is convex on</u> $[a,b]$.

<u>Remark.</u> This is a generalization of the fact that $\overline{D}_2(f,x) \geq 0$ implies f is convex.

<u>Proof.</u> Here again, our argument is the parabola technique. If f is not convex on $[a,b]$ then for some $a \leq x_o < \xi \leq y_o \leq b$. The point $(\xi,f(\xi))$ lies

above the line connecting $(x_0, f(x_0))$ to $(y_0, f(y_0))$. Thus, there is a point $\eta \in (x_0, y_0)$ and a parabola $Q(x) = \alpha(x-\eta)^2 + \beta(x-\eta) + f(\eta)$, with $\alpha < 0$, which lies above f on $[x_0, y_0]$. The existence of such a parabola is established as we have argued in Lemma 5.3. Let $a \leq c \leq x_0$ be the largest number such that $Q(c) = f(c)$. If no such points exist, we let $c = a$. Similarly, let $y_0 \leq d \leq b$ be the smallest number such that $Q(d) = f(d)$. Now define

$$g(t) = \begin{cases} 0 & \text{if } t \in [c,d] \\ f(t) - Q(t) & \text{if } t \notin [c,d] \end{cases}$$

then

$$f(t) \leq Q(t) + g(t) \quad \text{on} \quad [a,b]$$

and so

$$L_n(f,\eta) \leq L_n(Q,\eta) + o(\mu_n(\eta))$$

where we have used (5.5.1). This gives

$$\overline{\lim_{n\to\infty}} (\mu_n(\eta))^{-1}[L_n(f,\eta) - f(\eta)] \leq \overline{\lim_{n\to\infty}} (\mu_n(\eta))^{-1}[L_n(Q,\eta) - Q(\eta)] = 2\alpha < 0$$

where now we have used (5.5.3) to disregard the linear function. This is the desired contradiction.

We will need one more lemma which is in the spirit of the estimates given in Chapter 2 .

Lemma 5.5. *Let* (L_n) *be a sequence of positive operators mapping* $C[a,b]$ *into* $C[a,b]$ *and satisfying* (5.5.3) *at each point of* (a,b). *If* f *is a continuous function on* $[a,b]$ *and*

$$F(x) = \int_a^x f(t)\, dt \quad .$$

<u>Then for</u> $x_0 \in [a,b]$

$$\underline{D}(f,x_0) \leq \overline{\lim_{n\to\infty}}\, (\mu_n(x_0))^{-1}[L_n(F,x_0) - F(x_0)] \leq \overline{D}(f,x_0)$$

<u>where</u> $\underline{D}(f,x_0)$ <u>and</u> $\overline{D}(f,x_0)$ <u>are the lower and upper derivatives of</u> f <u>at</u> x_0 , <u>respectively</u>.

<u>Proof:</u> We will prove the inequality on the right. The left hand inequality is proved in the same way. First we note that

$$F(x) - F(x_0) - f(x_0)(x-x_0) = \int_{x_0}^{x} [f(t) - f(x_0)]dt \qquad (5.7.8).$$

If $\overline{D}(f,x_0)$ is $+\infty$, we have nothing to prove. In the other case let $d > \overline{D}(f,x_0)$. Then, there is a δ , such that $|t-x_0| < \delta$ implies

$$\frac{f(t) - f(x_0)}{(t-x_0)} \leq d \quad .$$

From this, it follows after integration that

$$\int_{x_0}^{x} [f(t) - f(x_0)]dt \leq \frac{1}{2}d\,(x-x_0)^2$$

provided $|x-x_0| < \delta$. Using this in (5.7.8), we see that

$$F(x) - F(x_0) - f(x_0)(x-x_0) \leq \frac{1}{2}d\,(x-x_0)^2 + R(x) \quad , \quad x \in [a,b] \qquad (5.7.9)$$

where R is a suitable continuous function which we can take to be 0 on

$[x_o - \frac{\delta}{2}, x_o + \frac{\delta}{2}]$.

Applying the operator L_n to both sides of (5.7.9), and using (5.5.1), we find

$$(\mu_n(x_o))^{-1} [L_n(F(x) - F(x_o) - f(x_o)(x-x_o), x_o)] \leq d + o_{x_o}(1) \qquad (5.7.10).$$

For each linear function ℓ ,

$$(\mu_n(x_o))^{-1} [L_n(\ell, x_o) - \ell(x_o)] = o_{x_o}(1)$$

by our hypothesis (5.5.3). This shows that (5.7.10) can be rewritten as

$$(\mu_n(x_o))^{-1} [L_n(F, x_o) - F(x_o)] \leq d + o_{x_o}(1) \quad .$$

The result then follows by taking $\overline{\lim_{n\to\infty}}$ and realizing that d was arbitrary.

We can now prove a pointwise "O" theorem due to H. Berens[4].

<u>Theorem 5.6.</u> <u>Let</u> (L_n) <u>be a sequence of positive operators which map</u> $C[a,b]$ <u>into</u> $C[a,b]$ <u>and satisfy</u> (5.5.3) <u>at each point of</u> (a,b) . <u>Let</u> $f \in C[a,b]$ <u>and suppose</u> g <u>is a function in</u> $L_1[a,b]$ <u>for which</u>

$$\underline{\lim}_{n\to\infty} (\mu_n(x))^{-1} [L_n(f,x) - f(x)] \leq g(x) \leq \overline{\lim_{n\to\infty}} (\mu_n(x))^{-1} [L_n(f,x) - f(x)] \quad ,$$

$$x \in (a,b) \qquad (5.7.11).$$

<u>Then there are constants</u> A <u>and</u> B <u>such that</u>

$$f(x) = Ax + B + \int_a^x \int_a^t g(u)du\, dt, \quad a \le x \le b \quad .$$

Proof: The proof is along the same lines as the proof of Lemma 5.3. We let m_n, m_n^*, M_n, M_n^* and H have the same meaning as they did in the proof of Lemma 5.3. We then have for each $k = 1,2,\ldots$,

$$\overline{\lim_{n\to\infty}} (\mu_n(x))^{-1} [L_n(f - m_k^*,x) - (f - m_k^*)(x)]$$

$$\ge g(x) - \overline{\lim_{n\to\infty}} (\mu_n(x))^{-1} [L_n(m_k^*,x) - m_k^*(x)]$$

$$\ge g(x) - \overline{D}(m_k,x) \ge 0 \quad .$$

Where for the first inequality, we used our hypothesis (5.7.11) about the relation of f and g. For the second inequality, we used Lemma 5.5. The last inequality is because m_k is a minor function for g. This shows, by Lemma 5.4, that $f - m_k^*$ is convex for each k. Taking a limit as $(k\to\infty)$, we see that H is convex. Arguing in a similar way for $f - M_k^*$, we find H is concave. Hence, H is linear and the theorem is proved.

Some remarks are in order. Theorem 5.6 contains Theorem 5.1 (when the asymptotic condition holds) by taking $g(x) = 0$ $x \in [a,b]$. Also, Theorem 5.6 contains Theorem 5.4, by taking for example, $g(x) = \overline{\lim_{n\to\infty}} (\mu_n(x))^{-1}[L_n(f,x)-f(x)]$. Since then $|g(x)| \le M$, so that the conclusion of Theorem 5.6 gives that

$$|f''(x)| \le M \quad \text{a.e. or} \quad f' \in \mathrm{Lip}\,(1,M) \quad .$$

5.8. More General Pointwise Saturation Theorems. We will show in this section how the preceding pointwise saturation results can be extended to include cases when the trivial class $T(L_n)$ is not necessarily the class of linear functions. The roles of the functions $1,t,t^2$ will be replaced by the more general triple of functions $1,u_1,u_2$ which form an extended complete Chebyshev system (see Section 1.2). We can choose u_1 and u_2 in such a way that

$$\begin{aligned} u_1(t) &= \int_a^t w_1(x)\,dx \\ W_2(t) &= \int_a^t w_2(x)\,dx \qquad\qquad (5.8.1) \\ u_2(t) &= \int_a^t W_2(x)\,w_1(x)\,dx \qquad a \le t \le b \end{aligned}$$

where w_1 and w_2 are both strictly positive on $[a,b]$, with w_2 continuous and w_1 continuously differentiable on $[a,b]$.

In order to obtain pointwise saturation theorems, we need to assume something about the behavior of (L_n). We do this in the form of an asymptotic condition which is the analogue of (5.5.3). If f is differentiable on $[a,b]$, let

$$D_{(w_1)}(f) = \frac{f'}{w_1} \quad , \quad D_{(w_2)}(f) = \frac{f'}{w_2} \qquad (5.8.2).$$

If f is twice continuously differentiable the composition operator $D_{(w_2,w_1)} = D_{(w_2)} \circ D_{(w_1)}$ can be applied to f. The asymptotic condition will be

<u>5.8.3</u>. <u>There exists a sequence</u> $(\mu_n(x))$ <u>of positive functions on</u> (a,b) <u>such that for each</u> $f \in C^{(2)}[a,b]$ <u>and</u> $x \in (a,b)$

$$\lim_{n\to\infty} (\mu_n(x))^{-1} (L_n(f,x) - f(x)) = D_{(w_2,w_1)}(f,x) = \left[\frac{1}{w_2}\left(\frac{f'}{w_1}\right)'\right](x)$$

In this form, we see that (5.8.3) is a generalization of the asymptotic condition (5.5.3). Since $D_{(w_2,w_1)}(1) = D_{(w_2,w_1)}(u_1) = 0$, (5.8.3) gives that

$$\text{if} \quad u \in sp(1,u_1)\text{ , then } L_n(u,x) - u(x) = o_x(\mu_n(x)),\ x \in (a,b) \qquad (5.8.4).$$

Also, for each $x \in (a,b)$, $\mu_n(x) = L_n(u_2,x) - u_2(x) + o_x(\mu_n(x))$. Thus, for each $x \in (a,b)$, $L_n(u_2,x) - u_2(x)$ is eventually positive. It will be convenient to require that $\mu_n(x)$ be defined by

$$\mu_n(x) = L_n(u_2,x) - u_2(x) \qquad (5.8.5).$$

Our approach to obtaining new pointwise theorems will be to reproduce the arguments in 5.7 in the more general framework given above. The essential arguments in 5.7 were convexity arguments. When $f \in C[a,b]$, we say f is convex with respect to $(1,u_1)$, if for each pair of points α and β in $[a,b]$, the function $u \in sp(1,u_1)$ which interpolates f at α and β lies above f on $[\alpha,\beta]$. Since $(1,u_1)$ is a Chebyshev system the function u is unique and in fact given by

$$u(x) = \frac{u_1(x) - u_1(\alpha)}{u_1(\beta) - u_1(\alpha)} f(\beta) + \frac{u_1(\beta) - u_1(x)}{u_1(\beta) - u_1(\alpha)} f(\alpha) \qquad (5.8.6).$$

Analogously, a function f is said to be concave with respect to $(1,u_1)$ if for each $\alpha,\beta \in [a,b]$ the corresponding function u lies below f on $[\alpha,\beta]$. Clearly, if f is both convex and concave with respect to $(1,u_1)$ then $f \in sp(1,u_1)$. We then say f is linear.

If f is convex with respect to $(1,u_1)$ and $\eta \in (a,b)$ then Z. Ziegler[1] has shown that there is a function $u \in sp(1,u_1)$ such that

$$u(x) \leq f(x) \qquad x \in [a,b]$$

$$u(\eta) = f(\eta) \qquad .$$

The following proof of this result was kindly communicated by H. Berens.

Lemma 5.6. *If* f *is convex with respect to* $(1,u_1)$ *and* $\eta \in (a,b)$ *then there is a function* $u \in sp(1,u_1)$ *such that*

$$u(x) \leq f(x) \qquad x \in [a,b]$$

and

$$u(\eta) = f(\eta) \qquad .$$

Proof: We first show that f has left $D^-_{u_1}(f,x)$ and right hand $D^+_{u_1}(f,x)$ derivatives with respect to u_1 which are finite at each point $x \in (a,b)$. If $\alpha < c < \beta$, then the function u in (5.8.6) satisfies

$$u(c) \geq f(c) \quad .$$

After some algebraic manipulation we find

$$\frac{f(c) - f(\alpha)}{u_1(c)-u_1(\alpha)} \leq \frac{f(\beta) - f(c)}{u_1(\beta)-u_1(c)} \qquad (5.8.7).$$

Note that $u_1(\alpha) < u_1(c) < u_1(\beta)$.

Thus, if $\alpha < c < d < \beta$, we find

$$\frac{f(c) - f(\alpha)}{u_1(c) - u_1(\alpha)} \leq \frac{f(d) - f(c)}{u_1(d) - u_1(c)} \leq \frac{f(\beta) - f(d)}{u_1(\beta) - u_1(d)} \quad .$$

Taking limits as $d \to c$, we deduce that

$$\frac{f(c) - f(\alpha)}{u_1(c) - u_1(\alpha)} \le \underline{D}^+_{u_1}(f,c) \le \overline{D}^+_{u_1}(f,c) \le \frac{f(\beta) - f(c)}{u_1(\beta) - u_1(c)} .$$

Now, we take a $\underline{\lim}$ as $\beta \to c$ to find

$$\underline{D}^+_{u_1}(f,c) \le \overline{D}^+_{u_1}(f,c) \le \underline{D}^+_{u_1}(f,c)$$

which shows that $D^+_{u_1}(f,c)$ exists and is finite. In a similar way, one can show $D^-_{u_1}(f,c)$ exists and is finite.

Now, let

$$u(x) = f(\eta) + D^+_{u_1}(f,\eta)\,(u_1(x) - u_1(\eta)) .$$

We will show u satisfies the conclusion of the lemma. If $a \le x < \eta$ then think of x as α and c as η in (5.8.7) and take a limit as $\beta \to \eta^+$ to find

$$(u_1(\eta) - u_1(x))\, D^+_{u_1}(f,\eta) \ge f(\eta) - f(x)$$

which is desired.

For $\eta < x \le b$, we again use (5.8.7), but now take $\alpha = \eta$, $\beta = x$ and $c < x$. Taking a limit as $c \to \eta$ we find

$$\frac{f(x) - f(\eta)}{u_1(x) - u_1(\eta)} \ge D^+_{u_1}(f,\eta)$$

which gives the desired result.

We begin toward our saturation theorems with the analogue of the parabola technique.

Lemma 5.7. Let $f \in C[\alpha,\beta]$ with $f(\alpha) = f(\beta) = 0$. If there is a point $x_0 \in (\alpha,\beta)$ with $f(x_0) > 0$, then there is a function u of the form

$$u(x) = c_2\, u_2(x) + c_0$$

with $c_2 < 0$, such that $u(x) \geq f(x)$ on $[\alpha,\beta]$, and $u(\eta) = f(\eta)$ for some $\eta \in (\alpha,\beta)$.

Proof: Let $M = 2||f||\,[\alpha,\beta]$. For $c_2 < 0$ and suitably small the function $c_2\, u_2 + M$ lies above f on $[\alpha,\beta]$ and the

$$\inf_{\alpha \leq x \leq \beta} (c_2 u_2(x) + M - f(x)) = m$$

is attained at some point $\eta \in (\alpha,\beta)$. This is because $M - f(\alpha) = M - f(\beta) > M - f(x_0)$. The function $c_2\, u_2 + (M-m)$ is the desired function.

The following lemma, which is essentially contained in Z. Ziegler$_1$, shows the relation between convexity and approximation by (L_n).

Lemma 5.8. Let (L_n) be a sequence of positive linear operators which satisfy (5.8.3). If $f \in C[a,b]$, then f is convex with respect to $(1,u_1)$ on $[a,b]$ if and only if

$$\overline{\lim_{n\to\infty}}\, (\mu_n(x))^{-1}\, (L_n(f,x) - f(x)) \geq 0 \quad , \quad x \in (a,b) \qquad (5.8.8).$$

Proof: First suppose f is convex and $\eta \in (a,b)$. Let u be a function in $sp(1,u_1)$ which satisfies Lemma 5.6. Then

$$0 = \lim_{n\to\infty} (\mu_n(\eta))^{-1} (L_n(u,\eta) - u(\eta)) \leq \overline{\lim_{n\to\infty}} (\mu_n(\eta))^{-1} (L_n(f,\eta) - f(\eta))$$

which shows (5.8.8) .

Conversely, suppose f satisfies (5.8.8) . If f is not convex then there are points α,β in $[a,b]$ such that the function $u \in sp(1,u_1)$ which interpolates f at α and β lies below f at some point $x_o \in (\alpha,\beta)$. That is to say,

$$f(x_o) - u(x_o) > 0$$

while

$$f(\alpha) - u(\alpha) = f(\beta) - u(\beta) = 0 \quad .$$

Thus, from Lemma 5.7, there is a function of the form $c_2u_2 + c_o$ with $c_2 < 0$ and a point $\eta \in (\alpha,\beta)$ such that

$$c_2u_2(x) + c_o \geq f(x) - u(x) \qquad x \in (\alpha,\beta)$$

and

$$c_2u_2(\eta) + c_o = f(\eta) - u(\eta) \quad .$$

Take a smaller interval $\alpha < \alpha' < \eta < \beta' < \beta$ and let g be a non-negative twice continuously differentiable function which vanishes on $[\alpha',\beta']$ and satisfies

$$g(x) \geq f(x) - u(x) - c_2u_2(x) - c_o \qquad x \notin [\alpha,\beta] \quad .$$

Then,

$$f(x) \leq g(x) + u(x) + c_2 u_2(x) + c_o \qquad x \in [a,b] \quad .$$

Therefore,

$$L_n(f,\eta) - f(\eta) \leq c_2(L_n(u_2,\eta) - u_2(\eta)) + o(\mu_n(\eta))$$

$$= c_2 \, \mu_n(\eta) + o(\mu_n(\eta))$$

where to get the o term we have used (5.8.4) on $u + c_o$ and (5.8.3) on g since $D_{(w_2,w_1)}(g,\eta) = 0$. Since $c_2 < 0$, we have a contradiction to (5.8.8) and the lemma is proved.

Now, we can prove a "o" saturation theorem.

Theorem 5.7. *Let (L_n) be a sequence of positive linear operators which satisfies the asymptotic condition* (5.8.3). *If f is in $C[a,b]$ then*

$$L_n(f,x) - f(x) = o_x(\mu_n(x)) \quad , \quad x \in (a,b) \qquad (5.8.9)$$

if and only if f is linear with respect to $(1,u_1)$.

Proof: If f is linear we have already noted that (5.8.9) holds. If f satisfies (5.8.9), then by Lemma 5.8, f is both convex and concave on $[a,b]$, and hence f is linear.

To obtain a "O" saturation theorem, we will need a direct estimate in the flavor of Lemma 5.5.

Lemma 5.9. *Let (L_n) be a sequence of positive linear operators which satisfies the asymptotic condition (5.8.3). If $f \in C[a,b]$ and*

$$F(x) = \int_a^x f(t)\, w_1(t)\, dt$$

then for each $x_o \in (a,b)$

$$\underline{D}_{(w_2)}(f,x_o) \le \overline{\lim_{n\to\infty}} (\mu_n(x_o))^{-1}(L_n(F,x_o)-F(x_o)) \le \overline{D}_{(w_2)}(f,x_o) \quad (5.8.10).$$

Remark: $\overline{D}_{(w_2)}(f,x_o)$ is defined by $\overline{\lim_{t\to x_o}} \dfrac{f(t) - f(x_o)}{W_2(t) - W_2(x_o)}$, with

$$W_2(t) = \int_a^t w_2(u)\, du \quad .$$

Proof: We will only prove the right hand side of (5.8.10) . The other half is proved similarily. Let $x_o \in (a,b)$, if $\overline{D}_{(w_2)}(f,x_o) = \infty$ we have nothing to prove. Suppose $\overline{D}_{(w_2)}(f,x_o) < d$, where d is finite. It is enough to show that

$$\overline{\lim_{n\to\infty}} (\mu_n(x_o))^{-1}(L_n(F,x_o) - F(x_o)) < d \quad (5.8.11).$$

To show (5.8.11), we write

$$F(t) - F(x_o) = \int_{x_o}^t f(s)\, w_1(s)\, ds \quad .$$

Now, for some $\delta > 0$, whenever $|t-x_o| < \delta$, we have

$$\frac{f(t) - f(x_o)}{W_2(t) - W_2(x_o)} < d \quad .$$

So that choosing δ smaller if necessary, we have

$$\frac{f(t) - f(x_o)}{(t-x_o)w_2(x_o)} < d \quad (5.8.12)$$

where in (5.8.12) we have used the fact that $W_2'(x_o) = w_2(x_o)$ and

$W_2(t) - W_2(x_o) = w_2(x_o)(t-x_o) + o(t-x_o)$. Multiplying by w_1 and integrating, we find

$$\int_{x_o}^{t} (f(s) - f(x_o))\, w_1(s)\, ds < d\, w_2(x_o) \int_{x_o}^{t} (s-x_o)\, w_1(s)\, ds$$

or in other words,

$$F(t) - F(x_o) - f(x_o)(u_1(t) - u_1(x_o)) < dw_2(x_o) \int_{x_o}^{t} (s-x_o)w_1(s)\, ds \quad (5.8.13).$$

Now, we apply L_n to both sides of (5.8.13), divide by $\mu_n(x_o)$, and take a $\overline{\lim}_{n\to\infty}$ and find

$$\overline{\lim}_{n\to\infty} (\mu_n(x_o))^{-1} (L_n(F,x_o) - F(x_o)) < d$$

where we have used (5.8.4) to ignore the terms with u_1 , and we have used the asymptotic condition on the right hand side of (5.8.13) while noting that

$$D_{(w_2,w_1)} \left(\int_{x_o}^{t} (s-x_o)\, w_1(s)\, ds, x_o\right) = \frac{1}{w_2(x_o)} \ .$$

This proves the lemma.

We can now give a pointwise "O" theorem for (L_n) due to H. Berens[4].

<u>Theorem 5.8.</u> <u>Let</u> (L_n) <u>be a sequence of positive linear operators which satisfies the asymptotic condition</u> (5.8.3). <u>If</u> $f \in C[a,b]$ <u>and</u>

$$\underline{\lim}_{n\to\infty} (\mu_n(x))^{-1} (L_n(f,x) - f(x)) \le g(x) \le \overline{\lim}_{n\to\infty} (\mu_n(x))^{-1}(L_n(f,x)-f(x)) , \ x \in (a,b)$$

(5.8.14)

where g *is finite valued and in* $L_1[a,b]$ *, then there are constants* c_0 *and* c_1 *such that*

$$f(x) = c_0 + c_1 u_1(x) + \int_a^x w_1(t) \int_a^t g(s)\, w_2(s)\, ds\, dt \quad , \quad a \leq x \leq b \quad .$$

Proof: The proof parallels that of Theorem 5.6. Let m_n and M_n be minor and major functions of g with respect to w_2 (i.e. $\overline{D}_{(w_2)}(m_n,x) \leq g(x) \leq \underline{D}_{(w_2)}(M_n,x)$, $x \in (a,b)$) , which satisfy

$$|M_n(x) - \int_a^x w_2(t)\, g(t)\, dt| < \frac{1}{n}.$$

and

$$|m_n(x) - \int_a^x w_2(t)\, g(t)\, dt| < \frac{1}{n} \quad , \quad a \leq x \leq b \quad .$$

Let

$$m_n^*(x) = \int_a^x m_n(t)\, w_1(t) dt \quad , \quad M_n^*(x) = \int_a^x M_n(t)\, w_1(t) dt$$

and

$$H(x) = f(x) - \int_a^x w_1(t) \int_a^t g(s)\, w_2(s)\, ds\, dt \quad .$$

We want to show that H is linear with respect to $(1,u_1)$. We can show that H is both convex and concave. For example, to show that H is convex, we observe that for each k

$$\overline{\lim_{n\to\infty}} (\mu_n(x))^{-1} (L_n(f-m_k^*,x) - (f-m_k^*)(x))$$

$$\geq g(x) - \overline{\lim_{n\to\infty}} (\mu_n(x))^{-1} (L_n(m_k^*,x) - m_k^*(x))$$

$$\geq g(x) - \overline{D}_{(w_2)} (m_k,x) \geq 0 \tag{5.8.15}$$

where the first inequality is because of the relation of f and g, the second inequality is Lemma 5.9, and the final inequality is because m_k is a minor function for g. The inequality (5.8.15) gives by virtue of Lemma 5.8 that $f - m_k^*$ is convex for each k, taking a limit as $(k\to\infty)$ shows that H is convex. In a similar way we could show that H is concave which proves the theorem.

5.9. An Example. As an application of the pointwise saturation theorems developed in the preceding section, we consider the Bernstein power series operators which are defined for $f \in C[0,1]$ by

$$L_n(f,x) = (1-x)^n \sum_{k=0}^{\infty} f(\frac{k}{n+k}) \binom{n+k-1}{k} x^k \quad , \quad n \in N .$$

These operators were introduced by W. Meyer-König and K. Zeller. Owing to the identity

$$(1-x)^n \sum_{k=0}^{\infty} \binom{n+k-1}{k} x^k = 1 \quad , \quad 0 \leq x < 1 \tag{5.9.1}$$

we see that $L_n(1,x) = 1$, $0 \leq x < 1$, and $n \in N$.

We also wish to estimate $L_n(t,x)$ and $L_n(t^2,x)$. We will first develop some formulae. For any fixed positive integer m and $0 \leq x < 1$, we have

$$(1-x)^n \sum_{k=0}^{\infty} \frac{1}{(n+k+m)} \binom{n+k-1}{k} x^k = \frac{(1-x)^n}{(n-1)} \sum_{k=0}^{\infty} \binom{n+k-2}{k} \left(\frac{n+k-1}{n+k+m}\right) x^k$$

$$= \frac{1-x}{(n-1)} \left[(1-x)^{n-1} \sum_{k=0}^{\infty} \binom{n+k-2}{k} \left(1 - \frac{m+1}{n+k+m}\right) x^k\right]$$

$$= \frac{(1-x)}{(n-1)} + O(n^{-2}) \qquad (5.9.2)$$

where we have used (5.9.1) and the fact that $\frac{1}{(n+k+m)} \leq \frac{1}{n}$, $k \in N$. From (5.9.1), it easily follows that

$$(1-x)^n \sum_{k=0}^{\infty} \left(\frac{1}{n+k+m}\right)^2 \binom{n+k-1}{k} x^k = O(n^{-2}) \qquad (5.9.3).$$

A similar type estimate gives for $n \geq 1$,

$$x(1-x)^n \sum_{k=0}^{\infty} \frac{1}{k+1} \binom{n+k-1}{k} x^k = \frac{(1-x)^n}{(n-1)} \sum_{k=0}^{\infty} \binom{n+k-1}{k+1} x^{k+1}$$

$$= \frac{1-x}{n-1} - \frac{(1-x)^n}{n-1} = \frac{1-x}{n-1} + o_x(n^{-1})$$

$$0 < x < 1 \qquad (5.9.4).$$

Now, we will estimate $L_n(t,x)$. We have

$$L_n(t,x) = (1-x)^n \sum_{k=1}^{\infty} \left(\frac{k}{n+k}\right) \binom{n+k-1}{k} x^k$$

$$= (1-x)^n \sum_{k=1}^{\infty} \binom{n+k-2}{k-1} \left(1 - \frac{1}{n+k}\right) x^k$$

$$= x(1-x)^n \sum_{k=0}^{\infty} \binom{n+k-1}{k} \left(1 - \frac{1}{n+k+1}\right) x^k$$

$$= x\left(1 - \frac{(1-x)}{(n-1)}\right) + O(n^{-2}) \qquad (5.9.5)$$

where in the last equality we used (5.9.2).

To estimate $L_n(t^2,x)$, we argue as follows.

$$L_n(t^2,x) = (1-x)^n \sum_{k=1}^{\infty} \left(\frac{k}{n+k}\right)^2 \binom{n+k-1}{k} x^k$$

$$= x^2(1-x)^n \sum_{k=1}^{\infty} \binom{n+k-3}{k-2} \frac{k(n+k-1)(n+k-2)}{(k-1)(n+k)^2} x^{k-2} \qquad (5.9.6).$$

Note that the term for $k = 1$ in (5.9.6) is $nx(1-x)^n(n+1)^{-2} = o\,(n^{-1})$, $0 \leq x \leq 1$. So that, (5.9.6) can be rewritten as

$$L_n(t^2,x) = x^2(1-x)^n \sum_{k=2}^{\infty} \binom{n+k-3}{k-2}\left(1 + \frac{1}{k-1}\right)\left(1 - \frac{1}{n+k}\right)\left(1 - \frac{2}{n+k}\right) x^{k-2} + o\,(n^{-1})$$

$$= x^2 + x^2(1-x)^n \sum_{k=o}^{\infty} \binom{n+k-1}{k}\left[\frac{1}{k+1} - \frac{3}{n+k+2} + \frac{2}{(n+k+2)^2}\left(\frac{k+2}{k+1}\right) + \right.$$

$$\left. - \frac{3}{(k+1)(n+k+2)}\right]x^k.$$

If we break up the last sum into four parts corresponding to the terms $\frac{1}{k+1}$, $\frac{-3}{n+k+2}$, $\frac{2}{(n+k+2)^2}\left(\frac{k+2}{k+1}\right)$, $\frac{-3}{(k+1)(n+k+2)}$, then the third sum will be $o(n^{-1})$ because of (5.9.3). The last sum is smaller than or equal to

$$\frac{x(1-x)^n}{n} \sum_{k=o}^{\infty} \frac{1}{k+1} \binom{n+k-1}{k} x^k = o_x(n^{-2})$$

where we have used (5.9.4). The first two sums are given by (5.9.4) and (5.9.2), respectively, so that we get

$$L_n(t^2,x) = x^2 + \frac{x(1-x)}{n-1} - \frac{3x^2(1-x)}{n-1} + o_x(n^{-1}) = x^2 + \frac{x(1-x)(1-3x)}{n-1} + o_x(n^{-1})$$

$$(5.9.7).$$

We introduce the functions $w_1(x) = w_2(x) = (1-x)^{-2}$. The formulas (5.9.1), (5.9.5) and (5.9.7) can then be written as

$$\lim_{n\to\infty} \frac{2(1-x)^2}{x} n(L_n(e_i,x) - e_i = D_{(w_2,w_1)}(e_i,x) \qquad (5.9.8).$$

Of course (5.9.8) is our asymptotic condition. The verification that (5.9.8) holds for each twice differentiable f is completed by showing $L_n((t-x)^4,x) = o_x(n^{-1})$ and then using arguments similar to the way we proved that (5.5.2) implies (5.5.3) in Lemma 5.2. The operator $D_{(w_2,w_1)}$ vanishes for the functions 1 and $\frac{1}{1-x}$. Therefore, we get the following pointwise saturation theorem for the Meyer-Kōnig Zeller operators (L_n) .

Theorem 5.9. The Meyer-Kōnig-Zeller operators (L_n) are pointwise saturated on $[0,1]$ with order $\left(\frac{x}{2n(1-x)^2}\right)$. If $f \in C[0,1]$ and

$$L_n(f,x) - f(x) = o_x\left(\frac{x}{2n(1-x)^2}\right) \quad \text{for} \quad 0 < x < 1$$

then f is linear with respect to $(1,(1-x)^{-1})$. If $g \in L_1[0,1]$ and

$$\underline{\lim}_{n\to\infty} \frac{n(1-x)^2}{x}(L_n(f,x) - f(x)) \le g(x) \le \overline{\lim}_{n\to\infty} \frac{n(1-x)^2}{x}(L_n(f,x) - f(x)), 0 < x < 1$$

then

$$f(x) = c_0 + c_1(1-x)^{-1} + 2\int_0^x (1-t)^{-2} \int_0^t (1-s)^{-2} g(s)dsdt \quad , \quad 0 \le x < 1 .$$

5.10. Notes.

5.10.1 . There are at least two other useful ways of attacking the saturation problem in $C[a,b]$. One is the functional analytic technique used by G.G. Lorentz$_1$ in his treatment of the saturation of Bernstein polynomials. This technique was subsequently applied by Y. Suzuki$_1$, K. Ikeno and Y. Suzuki$_1$, and

Y. Suzuki and S. Watanabe$_1$.

A second more recent approach is described in the papers of R. Schnabl$_1$, C. Micchelli$_1$, and S. Karlin and Z. Ziegler$_1$. What is involved in the trigonometric convolution case can already be found in Butzer and Berens$_1$. The idea is to derive from (L_n) a continuous semi-group by taking limits of appropriate iterates of L_n, namely, $L_n^{k_n}$, where $k_n \phi_n \to t > 0$ ((ϕ_n) is the saturation order). This determines a continuous semi-group of operators $T(t)$, whose saturation properties are the same as (L_n). Saturation for a semi-group is well established in the works of H. Berens$_1$, and P.L. Butzer and H. Berens$_1$. The approach to pointwise saturation theorems developed in Sections 5.7 and 5.8 were first given for the norm case by G.G. Lorentz and L. Schumaker$_1$.

5.10.2. For $C[a,b]$, we have only been able to prove saturation theorems when an asymptotic condition like (5.8.3) holds. For convolution operators, the saturation theorems in Chapter 3 went much beyond this. The real difficulty in $C[a,b]$ is that due to the lack of periodicity, we can not work with translates of the function in place of the function. This means every point has a definite role in the interval and all saturation theorems are essentially local saturation theorems.

Regarding local saturation theorems, even in the convolution case, there are no general results (the convolution structure really does not help for local saturation), except for those given in this chapter. However, it is possible to give local saturation theorems for certain operators even when (5.8.3) does not hold. What is usually needed is some asymptotic formula for suitably smooth functions. However, this formula need not involve second order differential operators as in (5.8.3).

For example, G. Sunouchi$_2$ has determined the local saturation for Fejér and Riesz means of the Fourier series. The local saturation of the Fejér operators is based on the asymptotic formula

$$\lim_{n\to\infty} n(\sigma_n(f,x) - f(x)) = \tilde{f}'(x)$$

which holds whenever $\tilde{f}'$ is continuous.

5.10.3 . In DeVore [6], we have shown that if (L_n) is a sequence of positive convolution operators satisfying the asymptotic formula (3.5.1) then (L_n) satisfies the following pointwise "o" saturation theorem

$$L_n(f,x) - f(x) = o_x(1 - \rho_{1,n}) \quad , \quad x \in [-\pi,\pi]$$

if and only if f is constant.

When $\psi_k = k^2$ in (3.5.1), then the asymptotic condition

$$\lim(1 - \rho_{1,n})^{-1}(L_n(f,x) - f(x)) = f''(x)$$

holds for twice continuously differentiable f . Thus, in this case, Theorem 5.6 gives a companion pointwise "O" theorem. Also for certain special operators, H. Berens[2,3] has given pointwise "O" theorems. However, there are no general pointwise "O" theorems comparable to the norm case (Theorems 3.2 and 3.3).

CHAPTER 6

ALGEBRAIC POLYNOMIAL OPERATORS

6.1. Introduction. We turn our attention to examining the limitations on approximation by operators whose range is contained in the space of algebraic polynomials. We say that (L_n) is a sequence of algebraic polynomial operators, if each L_n maps $C[a,b]$ into $C[c,d]$, with $[c,d] \subseteq [a,b]$, and for each $n \in N$, $f \in C[a,b]$, $L_n(f)$ is a polynomial of degree $\leq \alpha n + \beta$ on $[c,d]$, where α and β are non-negative integers, which are fixed for the sequence (L_n). The results in this chapter will parallel to a large extent the results of Chapter 4. In some instances, our analysis will go beyond that of Chapter 4, particularily in Section 6.3. In this way, we will also be able to derive some new results for approximation by trigonometric polynomial operators.

Our first result due to Korovkin$_4$ is the algebraic analogue of Theorem 4.1 and shows that the degree of approximation by algebraic polynomial operators is also limited by (n^{-2}).

Theorem 6.1. *Let (L_n) be a sequence of positive, algebraic, polynomial operators mapping $C[a,b]$ into $C[c,d]$, with $[c,d] \subseteq [a,b]$. Then, for one of the test functions $e_i(x) = x^i$, $i = 0,1,2$, we have*

$$||e_i - L_n(e_i)|| \neq o(n^{-2}) \qquad (6.1.1).$$

Note: The norm, as usual, is taken in the image space and hence is the supremum over $[c,d]$.

Proof. This is easy to obtain from the results of Section 1.6. If y is the mid-point of the interval $[c,d]$, then as remarked in Section 1.6, the function $f(x) = |x-y|$ has degree of approximation on $[c,d]$: $E_n(f) > Cn^{-1}$, where C is some positive constant, depending only on $[c,d]$. Now if (6.1.1) were not true, then each function in Lip 1 on $[a,b]$ would be approximated with degree $o(n^{-1})$ on $[c,d]$. This is because of our fundamental estimate (2.3.2). Since f is such a function, we have the desired contradiction.

For trigonometric polynomial operators which were given by convolution, we were able to go well beyond (6.1.1) and actually show that for no function f can we have $||f - L_n(f)|| = o(n^{-2})$, unless f is constant. In the algebraic case, it can also happen that the function $e_1(x)$ is approximated with order $o(n^{-2})$ (see the examples given in Section 6.2). What is happening here can be made clearer, if we preview a result which will appear in Section 6.3. Namely, in Corollary 6.2, we will prove that if

$$||e_i - L_n(e_i)|| = O(n^{-2})$$

for $i = 0,1$, then

$$||e_2 - L_n(e_2)|| \neq o(n^{-2}) .$$

Thus, it is the function x^2 , which is the real limitation in (6.1.1). This motivates the following definition.

A sequence of positive, algebraic, polynomial operators is said to be optimal on $[c,d]$, if L_n maps $C[a,b]$ into $C[c,d]$ and for $i = 0,1$

$$||e_i - L_n(e_i)|| = o(n^{-2}) \tag{6.1.2}$$

while for e_2 ,

$$||e_2 - L_n(e_2)|| = O(n^{-2}) \quad . \tag{6.1.3}$$

Optimal sequences also have the property that they yield the Jackson estimates

$$||f - L_n(f)|| \leq C(\omega(f,n^{-1}) + n^{-2})$$

for continuous f and

$$||f - L_n(f)|| \leq C(\omega(f',n^{-1})n^{-1} + n^{-2})$$

when f is continuously differentiable. Of course, our definition of optimal is the analogue of the definition for trigonometric, polynomial operators.

In the next section, we will give a general method for constructing optimal sequences of algebraic, polynomial operators by using convolution with non-negative, algebraic polynomials. As we have previously remarked (Section 5.1), these kinds of operators cannot be used to approximate on all of $[a,b]$. They are only efficient on subintervals $[c,d]$ with $a < c < d < b$. However, our examples have the property that they are optimal on each such subinterval $[c,d]$.

6.2. Examples of optimal sequences. The purpose of this section is to construct optimal sequences of operators of convolution type. These operators will be of the form

$$L_n(f,x) = \int_{-\frac{1}{2}}^{\frac{1}{2}} f(t)\, P_n(t-x)\, dt$$

where P_n is a non-negative algebraic polynomial of degree $\alpha n + \beta$, with $\int_{-1}^{1} P_n(t)\, dt = 1$.

In Corollary 2.2, we showed that the degree of approximation of such a sequence is controlled by the second moments

$$\alpha_n^2 = \int_{-1}^{1} t^2\, P_n(t)\, dt \quad .$$

If we want the sequence to be optimal on $[-\delta,\delta]$, $\delta < \frac{1}{2}$, we need

$$\alpha_n = O(n^{-1}) \tag{6.2.1}$$

In order to guarantee that e_0 and e_1 are approximated with order $o(n^{-2})$, we need

$$\int_{S_\delta} P_n(t)\, dt = o(n^{-2}) \tag{6.2.2}$$

where $S_\delta = [-1,1]\setminus[-\frac{1}{2}+\delta\, ,\, \frac{1}{2}-\delta]$. If we want (6.2.2) to hold for each $\delta < \frac{1}{2}$, then this is equivalent to (see for comparison Lemma 5.2)

$$\int_{-1}^{1} t^4\, P_n(t)\, dt = o(n^{-2}) \tag{6.2.3}$$

In order to see if and how (6.2.1) can be achieved, we are led to consider the following extremal problem: Minimize $\int_{-1}^{1} t^2\, P(t)\, dt$, over all non-negative polynomials P of degree $\leq n$, with $\int_{-1}^{1} P(t)\, dt = 1$. This extremal problem is solved by the following **lemma** , which is the analogue of the Fejér-Korovkin theorem (Theorem 4.2).

Lemma 6.1. For $n \in N$, let

$$\lambda_n(t) = c_n \left(\frac{P_{2n}(t)}{(t^2 - x_{1,2n}^2)}\right)^2$$

where P_{2n} is the Legendre polynomial of degree $2n$, $x_{1,2n}$ its' smallest positive zero and c_n is a normalizing constant chosen so that

$$\int_{-1}^{1} \lambda_n(t)\, dt = 1 \ .$$

Then, λ_n is a non-negative algebraic polynomial of degree $4n-4$ with the property that if Q is any non-negative polynomial of degree $\leq 4n-3$, with $\int_{-1}^{1} Q(t)\, dt = 1$ then

$$x_{1,2n}^2 = \int_{-1}^{1} t^2 \lambda_n(t)\, dt \leq \int_{-1}^{1} t^2 Q(t)\, dt \ .$$

Remark: Since λ_n is a polynomial of degree $4n-4$, it also solves the extremal problem for polynomials of degree $4n-4$. The polynomial which solves the extremal problem for degrees $4n-2$ and $4n-1$ is

$$c_n(1-x^2)\left(\frac{P_{2n}^{(1,1)}(x)}{x^2 - \alpha_n^2}\right)^2$$

where $P_{2n}^{(1,1)}$ is the Jacobi polynomial of degree $2n$ which is orthogonal with respect to $(1-x^2)$ on $[-1,1]$ and α_n is its' smallest positive zero. The proof of this is the same as that given below for degree $4n-4$ except now, it is necessary to use the quadrature formula with nodes at $-1,1$ and the zeros of $P_{2n}^{(1,1)}$. This formula is exact for polynomials of degree $4n+1$ (see V. Krylov$_1$, pg. 172).

Proof: We will use the Gauss quadrature formula (Theorem 1.10) with nodes at the zeros of P_{2n}. We label these zeros as in Section 1.9,

$$-1 < x_{-n,2n} < \cdots < x_{-1,2n} < x_{1,2n} < \cdots < x_{n,2n} < 1 .$$

The quadrature formula is exact for polynomials of degree $4n-1$. Also, the zeros are symmetric about the origin $x_{-k,2n} = -x_{k,2n}$, the Cotes numbers $A_k(2n)$ of the quadrature formula are positive, and $A_k(2n) = A_{-k}(2n)$, for each $k = 1,2,\ldots,n$.

Now, let $Q(x) \geq 0$, $x \in [-1,1]$ and have degree $\leq 4n-3$, with $\int_{-1}^{1} Q(t)\,dt = 1$. Then, $t^2 Q(t)$ is a polynomial of degree $\leq 4n-1$, so that

$$\int_{-1}^{1} t^2 Q(t)\,dt = \sum_{k=1}^{n} x_{k,2n}^2 A_k(2n)\,(Q(x_{k,2n}) + Q(x_{-k,2n}))$$

$$\geq x_{1,2n}^2 \sum_{k=1}^{n} A_k(2n)\,(Q(x_{k,2n}) + Q(x_{-k,2n})) = x_{1,2n}^2 .$$

We have used the fact that $x_{1,2n}^2 \leq x_{k,2n}^2$, for $k = 1,2,\ldots,n$, and all the terms in the sum are non-negative. For the last equality, we realized that the sum equaled $\int_{-1}^{1} Q(t)\,dt = 1$.

The polynomial λ_n is of degree $4n-4$ and so

$$\int_{-1}^{1} t^2 \lambda_n(t)\, dt = \sum_{k=1}^{n} A_k(2n)\, x_{k,2n}^2\, (\lambda_n(x_{k,2n}) + \lambda_n(x_{-k,2n}))$$

$$= x_{1,2n}^2\, A_1(2n)\, (\lambda_n(x_{1,2n}) + \lambda_n(x_{-1,2n}))$$

$$= x_{1,2n}^2 \int_{-1}^{1} \lambda_n(t)\, dt = x_{1,2n}^2 \quad .$$

In the second equality, we used the fact that $\lambda_n(x_{k,2n}) = 0$, for $k \neq -1,1$. The lemma is proved.

What is of interest to us is how fast does $(x_{1,2n}^2)$ tend to 0 . We have an estimate for the zeros of P_{2n} in Theorem 1.12. In particular,

$$x_{1,2n} \leq \cos\left(\frac{n-\frac{1}{2}}{2n}\pi\right) \leq \frac{\pi}{4n} \quad .$$

Therefore, the polynomials λ_n do provide the estimate (6.2.1). It can be shown that they do not satisfy (6.2.2). However, the idea used in constructing λ_n is still useful. What we must do is divide out more zeros near the origin.

<u>Theorem 6.2.</u> <u>For</u> $n \in N$, <u>let</u>

$$\lambda_n^*(t) = c_n \left(\frac{P_{2n}(t)}{(t^2-x_{1,2n}^2)(t^2-x_{2,2n}^2)}\right)^2$$

<u>where</u> P_{2n} <u>is the Legendre polynomial of degree</u> $2n$, $x_{1,2n}$ <u>and</u> $x_{2,2n}$ <u>are its two smallest positive zeros and</u> c_n <u>is a normalizing constant chosen so that</u> $\int_{-1}^{1} \lambda_n^*(t)\, dt = 1$. <u>Then, the sequence of positive polynomial operators</u> Λ_n <u>given by the convolution formula</u>

$$\Lambda_n(f) = \int_{-\frac{1}{2}}^{\frac{1}{2}} f(t)\, \lambda_n^*(t-x)\, dt$$

is optimal on $[-\delta,\delta]$, _for each_ $\delta < \frac{1}{2}$.

Proof: We establish (6.2.1) as before. Namely, $\lambda_n^*(t)$ is of degree $4n-8$, so that

$$\int_{-1}^{1} t^2 \lambda_n^*(t)\, dt = \sum_{k=1}^{n} 2A_k(2n)\, x_{k,2n}^2\, \lambda_n^*(x_{k,2n})$$

$$\leq x_{2,2n}^2 \left(2A_1(2n)\, \lambda_n^*(x_{1,2n}) + 2A_2(2n)\, \lambda_n^*(x_{2,2n})\right)$$

$$= x_{2,2n}^2 = O(n^{-2})$$

where the last equality comes from Theorem 1.12. In a similar way, we verify (6.2.3). $t^4 \lambda_n^*(t)$ is of degree $4n-4$ and so

$$\int_{-1}^{1} t^4 \lambda_n^*(t)\, dt = \sum_{k=1}^{n} 2A_k(2n)\, x_{k,2n}^4\, \lambda_n^*(x_{k,2n}) \leq x_{2,2n}^4 = O(n^{-4}) .$$

R. Bojanic_1 has shown how the above method can be used to construct optimal sequences from a general sequence of orthogonal polynomials. To begin with, let w be any even function in $L_1[-1,1]$ which satisfies the following inequalities.

$$0 < m \leq w(t) \qquad \text{for} \qquad t \in [-1,1] \tag{6.2.4}$$

$$w(t) \leq M < +\infty \qquad \text{for} \qquad t \in [-\eta,\eta] \qquad \text{for some} \quad \eta < 1 \tag{6.2.5}$$

Let (Q_n) be the sequence of orthogonal polynomials (degree of $Q_n = n$) with respect to the weight function w on $[-1,1]$. We need to estimate the zeros of Q_{2n} which we do in the following lemmas. Again, our main vehicle is the Gauss quadrature formula with nodes at the zeros of Q_{2n}. We write the zeros of Q_{2n} in increasing order as:

$$-1 < x_{-n,2n} < x_{-n+1,2n} < \cdots < x_{-1,2n} < x_{1,2n} < \cdots < x_{n,2n} < 1 \quad (6.2.6).$$

These zeros are symmetric with respect to the origin.

<u>Lemma 6.2.</u> <u>Let</u> w <u>satisfy</u> (6.2.4) <u>and</u> (6.2.5) <u>then, for each fixed</u> k,

$$x_{k,2n} \to 0 \ (n \to \infty) \quad .$$

<u>Proof:</u> To see this, let P be any polynomial of degree n. Then, using the Gauss quadrature formula, we see that if $f \in C[-1,1]$,

$$\int_{-1}^{1} w(t)\, f(t)\, dt - \sum_{k=-n}^{n}{}' A_k(2n)\, f(x_{k,2n})$$

$$= \int_{-1}^{1} w(t)\, (f(t) - P(t))\, dt - \sum_{k=-n}^{n}{}' A_k(2n)\, [f(x_{k,2n}) - P(x_{k,2n})]$$

where $\sum'$ indicates that the term corresponding to $k = 0$ is omitted. Choosing P so that $||f - P|| = E_n(f)$, we have

$$\left|\int_{-1}^{1} w(t)\, f(t)\, dt - \sum_{k=-n}^{n}{}' A_k(2n)\, f(x_{k,2n})\right| \le 2E_n(f) \int_{-1}^{1} w(t)\, dt \to 0 \quad .$$

Now suppose $x_{k,2n} \to 0$, for $|k| < k_o$, but $x_{k_o,2n} \not\to 0$, $n \to \infty$. Let (n_j) be chosen so that $|x_{k_o,2n_j}| > \epsilon_o > 0$, for $j = 1,2,\ldots$ Let f be any continuous function on $[-1,1]$ which vanishes outside $(-\epsilon_o,\epsilon_0)$ and on $[-\frac{1}{2}\epsilon_o, \frac{1}{2}\epsilon_o]$. Then,

$$\sum_{k=-n_j}^{n_j}{}' A_{k,2n_j} f(x_{k,2n_j}) = 0$$

provided we take j large enough so that $|x_{k,2n_j}| < \frac{1}{2}\epsilon_o$, for $|k| < k_o$. Taking a limit as $j \to \infty$, we see that

$$\int_{-1}^{1} w(t)\, f(t)\, dt = 0 \quad .$$

But, this can only happen if

$$\int_{\frac{1}{2}\epsilon_o}^{\epsilon_o} w(t)\, dt = 0$$

which contradicts (6.2.4).

<u>Lemma 6.3.</u> <u>Let</u> w <u>satisfy</u> (6.2.4) <u>and</u> (6.2.5). <u>If</u> $x_{k,2n} \in [-\frac{1}{2}\eta, \frac{1}{2}\eta]$, <u>where</u> η <u>is given in</u> (6.2.5), <u>then the corresponding Cotes number</u> $A_k(2n)$ <u>satisfies the inequality</u>

$$A_k(2n) \leq C\, n^{-1}$$

<u>with</u> C <u>a constant independent of</u> n and k.

Proof: This is a lemma of P. Erdos and P. Turan[1]. We follow the proof given in G. Freud[1].

The idea is to construct a non-negative polynomial P of degree $\leq 4n-1$ such that

$$P(x_{k,2n}) = 1 \quad , \quad \text{and}$$

$$\int_{-1}^{1} w(t)\, P(t)\, dt \leq C\, n^{-1} \qquad (6.2.7)$$

with C a constant independent of n and k. Since, if such a P exists, we have

$$A_k(2n) \leq \sum_{j=-n}^{n}{}' A_j(2n)\, P(x_{j,2n}) = \int_{-1}^{1} w(t)\, P(t)\, dt \leq C\, n^{-1}$$

as desired.

Now to the construction of P. In the following C will always denote a constant independent of n and k. Let P_{2n-1} be the Legendre polynomial of degree $2n-1$, normalized so that

$$1 = P_{2n-1}(1) = ||P_{2n-1}||\ [-1,1] \qquad (6.2.8).$$

With this normalization, we have (see Davis[1], pg. 365)

$$P_{2n-1}(x) = 2^{-2n+1} \sum_{k=0}^{n-1} (-1)^k \binom{2n-1}{k} \binom{4n-2k-2}{2n-1} x^{2n-2k-1} \qquad (6.2.9)$$

and also

$$\int_{-1}^{1} (P_{2n-1}(x))^2 \, dx = 2(4n-1)^{-1} \tag{6.2.10}$$

We use Stirling's formula and (6.2.9) to find

$$|P'_{2n-1}(0)| = 2^{-2n+1} \binom{2n-1}{n-1} (2n) \geq C\, n^{\frac{1}{2}} \tag{6.2.11}$$

Let $Q(x) = C_n \left(\frac{P_{2n-1}(x)}{x}\right)^2$, where $C_n^{-1} = (P'_{2n-1}(0))^2$ so that $Q(0) = 1$. Because of (6.2.11), we have $C_n \leq C\, n^{-1}$ and so

$$|Q(x)| \leq C\, \delta^{-2}\, n^{-1} \qquad \text{for} \quad |x| \geq \delta \tag{6.2.12}$$

where we have also used (6.2.8). We want to observe that

$$\int_{-1}^{1} Q(t) \, dt \leq C\, n^{-1} \tag{6.2.13}$$

To see this, we use the Gauss quadrature formula based on the zeros of P_{2n-1}. Since Q vanishes at each of the zeros except the origin, we find

$$\int_{-1}^{1} Q(t) \, dt = \lambda_o^{(n)}$$

where $\lambda_o^{(n)}$ is the Cotes number corresponding to the point 0. Now the separation theorem (Theorem 1.11), gives that $\lambda_o^{(n)} \leq \alpha_n$ where α_n is the smallest positive zero of P_{2n-1}. We have an estimate for the size of α_n (Theorem 1.12), which shows

$$\int_{-1}^{1} Q(t) \, dt \leq \alpha_n \leq C\, n^{-1}$$

and this is (6.2.13).

Now we can take $P(t) = Q\left(\frac{t-x_{k,2n}}{1+\eta}\right)$, where η is given by (6.2.5). Note, we have assumed $x_{k,n} \in [-\frac{1}{2}\eta, \frac{1}{2}\eta]$. Let $S_\eta = [-1,1] \setminus [-\eta,\eta]$. If $t \in S_\eta$, then $|t - x_{k,2n}| \geq \frac{1}{2}\eta$, so that

$$\tfrac{1}{2}\eta(1+\eta)^{-1} \leq |(1+\eta)^{-1}(t - x_{k,2n})| \leq 1$$

and so using (6.2.12) we find

$$|P(t)| \leq C\, n^{-1} \quad , \quad t \in S_\eta \quad .$$

Hence, using (6.2.5) , we find

$$\int_{-1}^{1} w(t)\, P(t)\, dt = \int_{-\eta}^{\eta} w(t)\, P(t)\, dt + \int_{S_\eta} w(t)\, P(t)\, dt$$

$$\leq M \int_{-\eta}^{\eta} P(t)\, dt + C\, n^{-1} \int_{-1}^{1} w(t)\, dt \leq (1+\eta) M \int_{-1}^{1} Q(t)\, dt + C\, n^{-1} \leq C\, n^{-1}$$

where in the last inequality we used (6.2.13). Note also that

$$|\eta \pm x_{k,2n}| \leq \tfrac{3}{2}\eta \leq 1 + \eta \ , \quad \eta \leq 1 \quad .$$

This was used in changing variables on the integral of P . This verifies (6.2.7) and the lemma is proved.

<u>Lemma 6.4.</u> <u>Let</u> w <u>satisfy conditions</u> (6.2.4) <u>and</u> (6.2.5). <u>Then, for each fixed</u> k_o , <u>we have</u>

$$|x_{k_o,2n}| \leq C\, n^{-1} \quad , \quad n = 1,2,\ldots$$

where C is a constant independent of n.

Proof: We can assume that k_o is positive because of the symmetry of the zeros. We will use the separation theorem (Theorem 1.11). Let $y_{k,2n}$ be defined as in Theorem 1.11. For n sufficiently large

$$m\, x_{k_o,2n} \leq \int_0^{x_{k_o,2n}} w(t)\, dt \leq \int_{y_{-1,2n}}^{y_{k_o,2n}} w(t)\, dt = \sum_{-1}^{k_o} A_k(2n) \leq C(k_o+1)n^{-1}$$

where C is the constant of Lemma 6.3. The first inequality follows from (6.2.4), the second because $y_{-1,2n} < 0$, and the separation theorem. We used Lemma 6.3 and the fact that $|x_{k_o,2n}| < \frac{1}{2}\eta$, for n sufficiently large because of Lemma 6.2.

Theorem 6.3. *Let w be a weight function which satisfies (6.2.4) and (6.2.5) and (Q_n) the sequence of orthogonal polynomials with respect to w. The zeros of Q_{2n} are labeled as in (6.2.6). If k_o is an integer ≥ 2, let*

$$\lambda_n(t) = c_n \left[\frac{Q_{2n}(t)}{(t^2 - x_{1,2n}^2) \cdots (t^2 - x_{k_o,2n}^2)} \right]^2$$

with c_n a normalizing constant chosen so that $\int_{-1}^{1} \lambda_n(t)\, dt = 1$. Then, the sequence of operators (Λ_n) given by

$$\Lambda_n(f,x) = \int_{-\frac{1}{2}}^{\frac{1}{2}} f(t)\, \lambda_n(t-x)\, dt \qquad (6.2.14)$$

is optimal on $[-\delta,\delta]$, for each $\delta < \frac{1}{2}$.

Proof. The proof is the same as that of Theorem 6.2. We have

$$m \int_{-1}^{1} t^2 \lambda_n(t)\, dt \le \int_{-1}^{1} w(t)\, t^2 \lambda_n(t)\, dt \le x^2_{k_o,n} \le C\, n^{-2} \qquad (6.2.15).$$

The first inequality is from (6.2.4). The second inequality is obtained in the same way we have estimated in Theorem 6.2. The last inequality is just Lemma 6.4. To obtain (6.2.3), we estimate the same way

$$m \int_{-1}^{1} t^4 \lambda_n(t)\, dt \le \int_{-1}^{1} w(t)\, t^4 \lambda_n(t)\, dt \le x^4_{k_o,n} \le C\, n^{-4} = o(n^{-2}) \qquad (6.2.16).$$

The saturation properties of the sequence of operators given by (6.2.14) is easy to obtain from our saturation theorems of Chapter 5.

Corollary 6.1. Let (Λ_n) be the sequence of operators given by (6.2.14). Then for each $\delta < \frac{1}{2}$, (Λ_n) is locally saturated on $[-\delta,\delta]$ with order (n^{-2}) trivial class $T(\Lambda_n) = \{\ell : \ell \text{ is linear on } [-\delta,\delta]\}$ and saturation class $S(\Lambda_n) = \{f : f' \in \text{Lip } 1 \text{ on } [-\delta,\delta]\}$.

Proof. This follows from Corollary 5.4, where (6.2.15) shows that the second moments of λ_n are $O(n^{-2})$. We know that the second moments are $\ge C\, n^{-2}$ because of Theorem 6.2. The estimate (6.2.16) shows that the fourth moments are "o" of the second moments.

6.3. Local estimates for positive polynomial operators. We have suggested in Section 6.1 the possibility of a "o" saturation theorem for optimal operators. We now want to develop techniques which will allow us to show that if (L_n) is optimal from $C[-1,1]$ into $C[-\delta,\delta]$, for each $\delta < 1$, then

$$||f - L_n(f)||\ [-\delta,\delta] = o(n^{-2}) \quad \text{for each} \quad \delta < 1 \qquad (6.3.1)$$

if and only if f is linear on $[-1,1]$. In order to motivate the analysis that we will give in this section, let us see what we could obtain using the techniques we have developed so far.

If we use the parabola technique developed in Chapter 5, we could show that in the case f is not linear, there is a quadratic polynomial $Q(x) = \alpha x^2 + \beta x + \gamma$, with $\alpha < 0$ such that

$$Q(x) \geq f(x) \quad \text{on} \quad [-1,1] \qquad (6.3.2)$$

and $Q(\eta) = f(\eta)$ for some point $\eta \in (-1,1)$. Here, in order to have the polynomial Q lie above f, we may have to use the function $-f$ in place of f. Applying the operator L_n to (6.3.2) at the point η, we find (in the same way we have argued in Theorem 5.1) that

$$L_n((t-\eta)^2, \eta) = o(n^{-2}) \quad .$$

Unfortunately, this does not give a contradiction since all we know so far via Theorem 6.1 is that $||L_n((t-x)^2,x)||\ [a,b] \neq o(n^{-2})$, for each non-degenerate interval $[a,b]$. The problem here is that we have no control over the point $\eta \in (-1,1)$ where the parabola touches the function. Or viewed in another way, we don't have enough information on the behavior of the function $\mu_n(x) = \frac{1}{2}L_n((t-x)^2,x)$. Recall that in Chapter 5, pg. 136, we gave a general saturation theorem for the saturation order μ_n. Now, we want to replace μ_n by n^{-2}.

It is possible to control somewhat the point where the estimate is taken, if we assume further, that f is twice continuously differentiable. In this case, we can show that at the point η where the parabola touches f we must have $f''(\eta) < 0$. Then using the continuity of f'' we can find an interval $I = (\eta-\delta, \eta+\delta)$ of η and a new constant $\alpha < 0$ such that for each $y \in I$ the parabola

$$Q_y = \alpha(x-y)^2 + f'(y)(x-y) + f(y) \geq f(x) \quad , \quad x \in [-1,1] \quad .$$

This would then give the estimate

$$||L_n(t-x)^2,x)|| \; [\eta-\delta \; , \; \eta+\delta] = o(n^{-2})$$

which contradicts Theorem 6.1. Thus, we can prove a "o" theorem under the additional assumption that f'' is continuous. But the continuity of f'' (in fact even the existence of f'') cannot be deduced directly from (6.3.1) since the corresponding inverse theorem for approximation by algebraic polynomials gives only that f' is smooth on each closed subinterval of $(-1,1)$ (see remarks following Theorem 1.7).

We are led finally to approximate f by a sequence (f_n) of twice continuously differentiable function in hopes of obtaining the result. When we do this we find that in general for the corresponding intervals $[\eta_n-\delta_n \; , \; \eta_n+\delta_n]$, we have $\delta_n \to 0$. Because of this we will need to estimate $||L_n((t-x^2),x)|| \; [\eta_n-\delta_n,\eta_n+\delta_n]$. This is the subject of this section. What we will show is that if $\delta_n \neq o(n^{-1})$, then

$$||L_n((t-x)^2,x)||\ [\eta_n-\delta_n\ ,\ \eta_n+\delta_n] \neq o(n^{-2})\quad ,$$

provided the points η_n stay away from the end points -1 and 1. After obtaining this estimate, we will in Section 6.4 obtain the "o" saturation theorem using the plan outlined above.

We know from Theorem 1.6 that approximation by algebraic polynomials can be made more efficient at the end points of the interval $[-1,1]$ while retaining the same precision on the whole interval. Our next result will show this is a unique property of the end points. We first establish a corresponding result for trigonometric approximation.

<u>Lemma 6.5.</u> <u>If</u> $M^* > 0$ <u>is any constant, then there are positive constants</u> C_1^* <u>and</u> C_2^*, <u>with the property that if</u> $\eta \in [-\pi,\pi]$, <u>and</u> $n \in N$, <u>then there exists a</u> 2π <u>periodic even function</u> $g_n(\eta,x)$ <u>in</u> Lip (1,1), <u>such that whenever</u> T <u>is a polynomial of degree</u> $\leq n$ <u>with</u>

$$||g_n(\eta,x) - T(x)||\ [-\pi,\pi] \leq M^* n^{-1} \tag{6.3.3}$$

<u>we have</u>

$$||g_n - T||\ [\eta - C_1^*\, n^{-1}\ ,\ \eta + C_1^*\, n^{-1}] \geq C_2^*\, n^{-1} \tag{6.3.4}$$.

<u>Proof:</u> We will use the representation for trigonometric polynomials introduced in Section 1.8. We can assume that $\eta \in [0,\pi]$, since the function g_n which we construct is going to be even. Let $x_k = (2k\pi)(3n)^{-1}$, for k an integer. Let k_0 be that integer for which

$$\eta \in [\frac{2\pi k_o}{3n}, \frac{2\pi(k_o+1)}{3n}] = I_o$$

and let y be the mid-point of I_o. If $|y-x_k| \le \pi$, we have from (1.8.3) that

$$|V_n(y-x_k)| \le \frac{\pi^2}{2n}(y-x_k)^{-2} = \frac{9n}{8(k-k_o-\frac{1}{2})^2} \qquad (6.3.5).$$

Consider the 2π periodic <u>even</u> function $g_n(\eta,x)$ which is zero on $[0,\pi]\setminus I_o$ and the roof function $\frac{\pi}{3n} - |x-y|$, $x \in I_o$. We shall show that $g_n(\eta,x)$ satisfies the conclusion of the lemma with C_1^* and C_2^* to be determined below. It is clear that $g_n(\eta,x) \in \mathrm{Lip}\,(1,1)$.

Suppose T is a trigonometric polynomial of degree $\le n$ satisfying (6.3.3) and <u>not</u> satisfying (6.3.4). Because of the periodicity of T and V_n, we can take the sum in (1.8.1) over all the x_k in $[y-\pi, y+\pi]$. Thus, from (6.3.5), we have

$$|T(y)| \le \frac{2}{3n} \sum_{2|x_k-y| \le C_1^* n^{-1}} |T(x_k)| \frac{9n}{8(k-k_o-\frac{1}{2})^2}$$

$$+ \frac{2}{3n} \sum_{2|x_k-y| > C_1^* n^{-1}} |T(x_k)| \frac{9n}{8(k-k_o-\frac{1}{2})^2} \qquad (6.3.6).$$

Now, $g_n(\eta,x_k) = 0$ for all k and if $2|y-x_k| \le C_1^* n^{-1}$, then $|\eta-x_k| \le C_1^* n^{-1}$ provided $3C_1^* \ge 4\pi$ which is a restriction we will impose on C_1^*. Hence, using (6.3.3) on the second sum in (6.3.6) and the assumption that (6.3.4) does not hold on the first sum, we find

$$|T(y)| \leq \frac{3C_2^*}{4n} \sum_{2|x_k-y|\leq C_1^* n^{-1}} (k-k_o-\tfrac{1}{2})^{-2} + \frac{3M^*}{4n} \sum_{2|x_k-y|\geq C_1^* n^{-1}} (k-k_o-\tfrac{1}{2})^{-2}$$

$$\leq \frac{3C_2^*}{2n} \sum_{k=0}^{\infty} (k-\tfrac{1}{2})^{-2} + \frac{3M^*}{2n} \sum_{4(k-\frac{1}{2})\pi>3C_1^*} (k-\tfrac{1}{2})^{-2} \qquad (6.3.7).$$

We now choose C_2^* so small that $C_2^* < \frac{4}{5}$ and

$$3C_2^* \sum_{k=0}^{\infty} (k-\tfrac{1}{2})^{-2} < \frac{1}{5}$$

and we choose C_1^* so large that $3C_1^* > 4\pi$ and

$$3M^* \sum_{4(k-\frac{1}{2})\pi>3C_1^*} (k-\tfrac{1}{2})^{-2} < \frac{1}{5} .$$

We then have by virtue of (6.3.7) that

$$|T(y)| < \frac{1}{10n} + \frac{1}{10n} = \frac{1}{5n} \qquad (6.3.8).$$

Since $g_n(\eta,y) = \frac{\pi}{3n}$, we see from (6.3.8) that

$$|g_n(\eta,y) - T(y)| \geq \frac{\pi}{3n} - \frac{1}{5n} \geq \frac{4}{5n} > C_2^* n^{-1}$$

the desired contradiction.

The result we want for the algebraic case is now given by the following lemma.

Lemma 6.6. Let $0<\delta<1$. If $M > 0$ is any constant, then there are positive constants C_1 and C_2 depending only on δ and M with the property that *if* $x_0 \in [-\delta,\delta]$ *and* $n \in N$, *then there exists a function* $f_n(x_0,x) \in Lip(1,1)$ *on* $[-1,1]$ *such that whenever* P *is an algebraic polynomial of degree* $\leq n$ *with*

$$||f_n - P||\ [-1,1] \leq M\,n^{-1} \qquad (6.3.9)$$

then, we must have

$$||f_n - P||\ [x_0 - C_1 n^{-1} , x_0 + C_1 n^{-1}] \geq C_2(1-x_0^2)^{\frac{1}{2}}\, n^{-1} \qquad (6.3.10).$$

Proof. The basic idea is to derive this result from Lemma 6.5 by means of the standard transformation $x = \cos\theta$. We will consider only $x_0 \in [0,\delta]$, the other case is handled similarily.

Let C_1^* and C_2^* be the constants of Lemma 6.5 corresponding to $M^* = 2(1-\delta^2)^{-\frac{1}{2}}M$. Furthermore, let $\eta = \cos^{-1}(x_0 - C_1^* n^{-1})$. We need only verify the lemma for n large enough so that η is defined. Set

$$h_n(x_0,x) = g_n(\eta,\cos^{-1}x)$$

where g_n is the function defined in the proof of Lemma 6.5. Now, g_n vanishes outside of $[\eta-2\pi(3n)^{-1} , \eta+2\pi(3n)^{-1}]$ and thus h_n will vanish if $|\cos^{-1}x - \eta| \geq 2\pi(3n)^{-1}$ and in particular when $|x - x_0 + C_1^* n^{-1}| \geq 2\pi(3n)^{-1}$. Here, we have used the fact on $[-1,1]$,$(\cos^{-1}x)' = -(1-x^2)^{-\frac{1}{2}} < -1$.

The function h_n is differentiable at all but three points and of course its' derivative is zero when $|x - x_0 + C_1^* n^{-1}| \geq 2\pi(3n)^{-1}$. In any case, when h_n is differentiable at x , we have

$$|h_n'(x_o,x)| = |g_n'(n,\cos^{-1}x)(1-x^2)^{-\frac{1}{2}}| \leq (1-x^2)^{-\frac{1}{2}} .$$

When $|x - x_o + C_1^* n^{-1}| \leq 2\pi(3n)^{-1}$, then

$$(1-x^2)^{-\frac{1}{2}} \leq (1 - (|x_o|+ \lambda_n)^2)^{-\frac{1}{2}}$$

where $\lambda_n = C_1^* n^{-1} + 2\pi(3n)^{-1}$. Since $0 \leq x_o \leq \delta$, for n sufficiently large, say $n \geq N$, with N independent of x_o

$$(1 - (|x_o|+ \lambda_n)^2)^{-\frac{1}{2}} \leq 2(1-x_o^2)^{-\frac{1}{2}} .$$

Thus, for $n \geq N$ the function

$$f_n(x_o,x) = \frac{1}{2}(1-x_o^2)^{\frac{1}{2}} h_n(x_o,x)$$

is in Lip (1,1) . We will now show that f_n satisfies the conclusions of the lemma where we have not yet specified C_1 or C_2 .

Suppose P is a polynomial of degree $\leq n$ and

$$||f_n - P|| \; [-1,1] \leq M n^{-1} .$$

Let $Q(x) = 2(1-x_o^2)^{-\frac{1}{2}} P(x)$, so that

$$||h_n - Q|| \; [-1,1] \leq 2M(1-x_o^2)^{-\frac{1}{2}} n^{-1} \leq M^* n^{-1}$$

or equivalently

$$||g_n(\theta) - Q(\cos\theta)||\ [-\pi,\pi] \le M^* n^{-1} .$$

From Lemma 6.5, it follows that

$$||g_n(\theta) - Q(\cos\theta)||\ [\eta - C_1^* n^{-1} , \eta + C_1^* n^{-1}] \ge C_2^* n^{-1} .$$

Since

$$[\eta - C_1^* n^{-1} , \eta + C_1^* n^{-1}] \subseteq [\cos^{-1} x_0 , \cos^{-1}(x_0 - 2C_1^* n^{-1})] ,$$

we have

$$||h_n - Q||\ [x_0 - 2C_1^* n^{-1} , x_0] \ge C_2^* n^{-1}$$

and so

$$||f_n - P||\ [x_0 - 2C_1^* n^{-1} , x_0 + 2C_1^* n^{-1}] \ge \tfrac{1}{2}C_2^*(1-x_0^2)^{\frac{1}{2}} n^{-1} \qquad (6.3.11).$$

Now, we specify $C_1 = 2C_1^*$ and $2C_2 < C_2^*$ so small that (6.3.11) will automatically hold for $n \le N$. The lemma is proved.

The following two theorems give local estimates for the degree of approximation by polynomial operators, the first for the trigonometric case, the second for the algebraic.

<u>Theorem 6.4.</u> <u>Let (L_n) be a sequence of trigonometric polynomial operators such that</u>

$$||g - L_n(g)|| = O(n^{-2})$$

for the three functions $g = 1$, $\cos x$, $\sin x$. *Then there are constants* C_1 *and* C_2 *such that for each* $x_o \in [\pi,\pi]$

$$||L_n(\sin^2(\tfrac{t-x}{2}) , x)|| \; [x_o - C_1 n^{-1} , x_o + C_1 n^{-1}] \geq C_2 n^{-2} .$$

Theorem 6.5. *Let* (L_n) *be a sequence of positive algebraic polynomial operators and let* $0 < \delta < a$. *If for* $i = 0,1,2$, we have

$$||e_i - L_n(e_i)|| \; [-a,a] = O(n^{-2})$$

then there are constants C_1 , C_2 *and* N *such that for any* $x_o \in [-\delta,\delta]$

$$||L_n((t-x)^2,x)|| \; [x_o - C_1 n^{-1} , x_o + C_1 n^{-1}] \geq C_2(a^2-x_o^2)n^{-2} \qquad , \qquad n \geq N .$$

Proof. The proofs of Theorems 6.4 and 6.5 follow from Lemmas 6.5 and 6.6 respectively. We will show Theorem 6.5. We can assume $a = 1$. The general case is obtained via the usual transformation of $[-a,a]$ into $[-1,1]$. For simplicity, let the range of L_n be in the space of polynomials of degree $\leq n$. From Theorem 2.3, it follows that for each $f \in Lip(1,1)$

$$||f - L_n(f)|| \; [-1,1] \leq C(||f||n^{-2} + n^{-1}) \qquad (6.3.12)$$

for a certain constant C which is independent of n and f . Choose $\overline{C_1}$ and $\overline{C_2}$ as in Lemma 6.6 corresponding to $\overline{M} = 2C$. Then, again using Theorem 2.3, we find that when $f \in Lip(1,1)$

$$||f - L_n(f)|| \; [x_o - \overline{C_1}\, n^{-1}, \; x_o + \overline{C_1}\, n^{-1}] \leq C(||f||n^{-2} + \mu_n^{\frac{1}{2}}) \qquad (6.3.13)$$

where

$$\mu_n = ||L_n((t-x)^2, x)|| \; [x_o - \overline{C_1}\, n^{-1}, \; x_o + \overline{C_1}\, n^{-1}] \; .$$

The constants C in (6.3.12) and (6.3.13) can be chosen to be the same.

The function $f_n(x_o, x)$ introduced in Lemma 6.6 is in $Lip(1,1)$ and has norm $\leq \frac{1}{2}(1-x_o^2)^{\frac{1}{2}} \leq 1$. Hence, from Lemma 6.6 we must have

$$\overline{C_2}(1-x_o^2)^{\frac{1}{2}}\, n^{-1} \leq C\, n^{-2} + C\, \mu_n^{\frac{1}{2}} \; .$$

Squaring both sides, we find

$$\mu_n \geq \overline{C_2}^{\,2}\, C^{-2}(1-x_o^2)n^{-2} - n^{-4} - 2n^{-2}\, \mu_n^{\frac{1}{2}} \; .$$

Now, we can let $C_2 = \frac{1}{2}\overline{C}_2^{\,2}\, C^{-2}$ and $C_1 = \overline{C_1}$ and realize that

$$n^{-2} + 2\mu_n^{\frac{1}{2}} \leq C_2(1-\delta^2) \leq C_2(1-x_o^2)$$

if n is sufficiently large, say $n \geq N$. Thus,

$$\mu_n \geq C_2(1-x_o^2)n^{-2} \quad , \quad n \geq N$$

which proves the theorem.

Corollary 6.2. *Let* (L_n) *be a sequence of positive operators on* $C[-a,a]$ *with* $0 < a$. *If for* $i = 0,1$ *we have*

$$||e_i - L_n(e_i)||\ [-a,a] = O(n^{-2})$$

then

$$||e_2 - L_n(e_2)||\ [-a,a] \neq o(n^{-2}) \quad .$$

Proof. We need only consider the case when $||e_2 - L_n(e_2)||\ [-a,a] = O(n^{-2})$. Let C_1 and C_2 be the constants from Theorem 6.5. Since $(t-x)^2 = e_2(t) -2x\, e_1(t) + x^2 e_0(t)$, we see that for n sufficiently large

$$C_2\ a^2 n^{-2} \leq ||L_n((t-x)^2,x)||\ [-C_1 n^{-1}\ ,\ C_1 n^{-1}]$$

$$\leq ||e_2 - L_n(e_2)||\ [-C_1 n^{-1}\ ,\ C_1 n^{-1}] + o(n^{-2}) \quad .$$

This first inequality comes from Theorem 6.5 and the second because $|x| \leq C_1\ n^{-1}$ on the interval where we take the norm. This proves the corollary.

6.4. *A "o" saturation theorem for optimal sequences.* At the beginning of Section 6.3, we outlined a plan for proving a "o" saturation theorem for optimal sequences. We will now supply the details of this plan. We had mentioned that we would approximate f by twice differentiable functions. The method of approximation is given in the following lemma.

<u>Lemma 6.7.</u> <u>Let</u> f <u>be a continuously differentiable function on</u> $[-1,1]$. <u>Extend</u> f <u>to be a continuous function on</u> $(-\infty,\infty)$ <u>by defining</u> f <u>to be constant on</u> $(-\infty,-1)$ <u>and</u> $(1,\infty)$. <u>Define</u>

$$f_n(x) = n^2 \int_{-\frac{1}{2}n^{-1}}^{\frac{1}{2}n^{-1}} \int_{-\frac{1}{2}n^{-1}}^{\frac{1}{2}n^{-1}} f(x+u+v)\,du\,dv \qquad n \in N \qquad (6.4.1).$$

<u>The sequence</u> (f_n) <u>converges uniformly to</u> f <u>on</u> $[-1,1]$. <u>If</u> $0<\delta<\delta'<1$, <u>there is an</u> N <u>such that for</u> $n \geq N$, <u>each</u> f_n <u>has a continuous third derivative on</u> $[-1,1]$,

$$||f_n^{(3)}||\ [-\delta,\delta] \leq n\,\varepsilon\,(n) \qquad (6.4.2)$$

<u>and</u>

$$||f' - f_n'||\ [-\delta,\delta] \leq \frac{\varepsilon(n)}{2n} \qquad (6.4.3)$$

<u>where</u> $\varepsilon(n) = \sup_{0\leq h\leq n^{-1}} h^{-1}||f'(x+h) + f'(x-h) - 2f'(x)||\ [-\delta',\delta']$.

<u>Proof.</u> The uniform convergence of (f_n) to f is clear. In order to verify (6.4.2), let F_1 be a primitive of f and F_2 a primitive of F_1 . Then

$$f_n(x) = n^2(F_2(x + \frac{1}{n}) + F_2(x - \frac{1}{n}) - 2F_2(x))$$

so that

$$f_n^{(3)}(x) = n^2[F_2^{(3)}(x + \frac{1}{n}) + F_2^{(3)}(x - \frac{1}{n}) - 2F_2^{(3)}(x)]$$

$$= n^2[f'(x + \frac{1}{n}) + f'(x - \frac{1}{n}) - 2f'(x)] \quad .$$

We need only take norms and use the definition of $\varepsilon(n)$ to obtain (6.4.2) for $n \geq N$ where N is chosen so that $N^{-1} < \delta' - \delta$.

To estimate (6.4.3) we write

$$f_n'(x) - f'(x) = n^2 \int_{-\frac{1}{2}n^{-1}}^{\frac{1}{2}n^{-1}} \int_{-\frac{1}{2}n^{-1}}^{\frac{1}{2}n^{-1}} [f'(x+u+v) - f'(x)]du\, dv$$

$$= \frac{n^2}{2} \int_{-\frac{1}{2}n^{-1}}^{\frac{1}{2}n^{-1}} \int_{-\frac{1}{2}n^{-1}}^{\frac{1}{2}n^{-1}} [f'(x+u+v) + f'(x-u-v) - 2f'(x)]du\, dv \quad .$$

This gives for $n \geq N$,

$$||f_n' - f'||\ [-\delta,\delta] \leq \frac{n^2}{2} \int_{-\frac{1}{2}n^{-1}}^{\frac{1}{2}n^{-1}} \int_{-\frac{1}{2}n^{-1}}^{\frac{1}{2}n^{-1}} \varepsilon\ (n)\ |u+v|\ du\, dv \leq \frac{\varepsilon(n)}{2n} \; .$$

The following lemma is rather technical and is best viewed as a refinement of the parabola technique introduced in Chapter 5.

Lemma 6.8. _Let_ f _be a continuous function on_ $[-1,1]$ _whose first derivative is smooth on_ $[-\delta,\delta]$ _for each_ $\delta < 1$ _and let_ f_n _be defined as in_ Lemma 6.7. _Suppose_ $f(-1) = f(1) = 0$ _and_ $f(x_0) > 0$ _for some_ $x_0 \in (-1,1)$. _If_ $C > 0$, _then there is an integer_ N, $\alpha < 0$, _and_ $0 < \delta_0 < 1$ _such that for each_ $n \geq N$ _we can find a point_ $x_n \in [-\delta_0,\delta_0]$ _for which the family of parabolas_

$$Q_n(x,y) = \alpha(x-y)^2 + f_n'(y)(x-y) + f_n(y)$$

satisfy

$$Q_n(x,y) \geq f_n(x) \quad \text{if} \quad -1 \leq x \leq 1 \quad \text{and} \quad |y-x_n| < C\, n^{-1} \; .$$

Remark. It will follow in the proof that δ_o depends only on f and hence not on C .

Proof. First choose δ_o sufficiently close to 1 so that

$$|f(x)| < \frac{1}{3}|f(x_o)| \qquad x \notin (-\delta_o,\delta_o) \quad .$$

Since (f_n) converges uniformly to f , we have for n sufficiently large, say $n \geq N_1$

$$|f_n(x)| \leq \frac{1}{3}|f_n(x_o)| \quad x \notin (-\delta_o,\delta_o) \quad .$$

Arguing as in the proof of Lemma 5.1, we can find a quadratic $R(x) = \alpha_1(x-x_o)^2 + \beta$, with $\alpha_1 < 0$ such that for each $n \geq N_1$

$$R(x) \geq f_n(x) \qquad x \in [-1,1] \quad ,$$

and $m_n = \inf\{R(x) - f_n(x) : x \in [-1,1]\}$ is attained on $[-\delta_o,\delta_o]$, say at the point x_n . The quadratics $R_n(x) = R(x) - m_n$ have the properties

$$R_n(x) = \alpha_1(x-x_n)^2 + f_n'(x_n)(x-x_n) + f_n(x_n) \qquad (6.4.4)$$

$$R_n(x) \geq f_n(x) \qquad x \in [-1,1] \qquad (6.4.5)$$

$$R_n(x_n) = f_n(x_n) \qquad (6.4.6).$$

The expansion (6.4.4) follows from the continuity of f_n and f_n' .

We will now show the above choices for x_n and δ_o work when we take $\alpha = \frac{1}{2}\alpha_1$. We want to show that the polynomials $Q_n(x,y) = \frac{1}{2}\alpha_1(x-y)^2 + f_n'(y)(x-y) + f_n(y)$ have the property

$$Q_n(x,y) \geq f_n(x) \quad , \quad x \in [-1,1] \quad , \quad |y-x_n| \leq C\, n^{-1} \tag{6.4.7}$$

provided n is sufficiently large. Let $\eta = \frac{1}{2}(1-\delta_o)$. We will have to consider two cases depending on whether $x \in [-1+\eta\ ,\ 1-\eta]$ or not. But, first we observe that if n is sufficiently large say $n \geq N_2 \geq N_1$, we have

$$|Q_n(x,y) - Q_n(x,x_n)| \leq |\tfrac{1}{2}\, \alpha_1 \eta^2| \quad , \quad |y-x_n| \leq C\, n^{-1} \tag{6.4.8}.$$

This follows from the equicontinuity of (f_n) and (f_n') .

Now, if $x \notin [-1+\eta\ ,\ 1-\eta]$, then $|x-x_n| \geq \eta$, so that for $n \geq N_2$

$$\begin{aligned} Q_n(x,y) &= Q_n(x,x_n) + (Q_n(x,y) - Q_n(x,x_n)) \\ &\geq \tfrac{1}{2}\, \alpha_1 (x-x_n)^2 + f_n'(x_n)(x-x_n) + f_n(x_n) + \tfrac{1}{2}\, \alpha_1\, \eta^2 \\ &\geq \alpha_1 (x-x_n)^2 + f_n'(x_n)(x-x_n) + f_n(x_n) = R_n(x) \geq f_n(x) \quad . \end{aligned}$$

In the first inequality, we used (6.4.8) (note that $\alpha_1 < 0$).

To see that (6.4.7) is valid for $x \in [-1+\eta\ ,\ 1-\eta]$ we introduce the functions

$$h_n(x,y) = \frac{f_n(x) - f_n(y) - f_n'(y)(x-y)}{(x-y)^2} \quad .$$

Clearly, it is sufficient to show that for n sufficiently large

$$h_n(x,y) \leq \tfrac{1}{2}\, \alpha_1 \tag{6.4.9}$$

provided $x \varepsilon [-1+\eta , 1-\eta]$ and $|y-x_n| < C n^{-1}$. Now suppose $x,y \varepsilon [-1 + \frac{1}{2}\eta , 1 - \frac{1}{2}\eta]$, and $\xi < \frac{1}{4}\eta$, then

$$|h_n(x+\xi , y+\xi) - h_n(x,y)| = \frac{1}{(x-y)^2} \int_y^x \int_t^{t+\xi} [f_n''(s) - f_n''(s+y-t)]ds\, dt$$

$$\leq \frac{\xi}{|x-y|} \omega(f_n'' , |x-y|) \leq n \varepsilon (n)\xi$$

provided n is sufficiently large, say $n \geq N_3 \geq N_2$. The last inequality comes from (6.4.2), with $\varepsilon(n)$ defined as in Lemma 6.7 by considering f on $[-1 + \frac{\eta}{4} , 1 - \frac{\eta}{4}]$. Thus, for $n \geq N_3$ and $\xi < C n^{-1}$

$$|h_n(x+\xi , y+\xi) - h_n(x,y)| \leq C \varepsilon (n) .$$

Finally, choosing n sufficiently large, say $n \geq N \geq N_3$, such that $C \varepsilon (n) < \frac{1}{2} |\alpha_1|$ and $C N^{-1} < \frac{1}{4}\eta$, we have for $|y-x_n| < C n^{-1}$ and $x \varepsilon [-1 + \frac{\eta}{2} , 1 - \frac{\eta}{2}]$ that

$$h_n(x + (y-x_n)),y) \leq h_n(x,x_n) + \frac{1}{2} |\alpha_1| \leq \alpha_1 + \frac{1}{2} |\alpha_1| \leq \frac{1}{2} \alpha_1$$

or

$$h_n(z,y) \leq \frac{\alpha_1}{2} , \quad |y-x_n| \leq C n^{-1} , \quad z \varepsilon [-1+\eta , 1-\eta] .$$

This is (6.4.9) and the lemma is proved.

We can now prove a "o" theorem for optimal sequences (see DeVore_2).

Theorem 6.6. *If* (L_n) *is a sequence of positive polynomial operators which is optimal on* $[-\delta,\delta]$ *for each* $\delta < 1$ *and if* $f \in C[-1,1]$ *with*

$$||f - L_n(f)||\ [-\delta,\delta] = o(n^{-2}) \qquad 0 < \delta < 1 \tag{6.4.10}$$

then f *is linear on* $[-1,1]$.

Proof: Without loss of generality, we can suppose that f vanishes at -1 and 1 . Since, if f does not, we can subtract the linear function which interpolates f at -1 and 1 and the resulting function will still satisfy (6.4.10).

Since $E_n(f) = o(n^{-2})$, f' is smooth on each subinterval $[-\delta,\delta]$, $\delta < 1$ (see Section 1.5). Suppose $f(x_0) > 0$ for some $x_0 \in (-1,1)$. Choose δ_0 as in Lemma 6.8 and let f_n be defined as in (6.4.1). We will also use Lemma 6.7 with $\delta = \delta_0$. From (6.4.3), it follows that $f'-f_n'$ is in $\text{Lip}(1,(2n)^{-1}\,\varepsilon\,(n))$ on $[-\delta_0,\delta_0]$. Therefore, using our fundamental estimates of Theorem 2.3, we find that

$$||f - f_n - L_n(f-f_n)||\ [-\delta_0,\delta_0] = o(n^{-2}) \quad .$$

We need only show that

$$||f_n - L_n(f_n)||\ [-\delta_0,\delta_0] \neq o(n^{-2}) \tag{6.4.11}$$

This will contradict (6.4.10) and show that $f(x) \leq 0$ on $(-1,1)$. Working with $-f$ in place of f , we find $f(x) \geq 0$ on $(-1,1)$ or $f = 0$, as desired. Now to establish (6.4.11), we will use Lemma 6.8 together with

Theorem 6.5. Let C_1 and C_2 be the constants of Theorem 6.5 corresponding to $\delta = \delta_o$ and $a = 1$. We use Lemma 6.8 with $C = C_1$. Recall our remark that δ_o of Lemma 6.8 does not depend on C so that we are okay in now prescribing $C = C_1$. The sequences $(||f_n||[-\delta_o,\delta_o])$ and $(||f'_n||[-\delta_o,\delta_o])$ are both bounded and so from Lemma 6.8 it follows that for each y such that $|x_n-y| \leq C\,n^{-1}$ we have

$$L_n(f_n,y) - f_n(y) \leq L_n(Q_n(x,y),y) - Q_n(y,y)$$

$$\leq \alpha L_n((x-y)^2,y) + 0\left(||e_1 - L_n(e_1)||\ [-\delta_o,\delta_o]\right)$$

$$+ 0\left(||e_0-L_n(e_0)||\ [-\delta_o,\delta_o]\right) \leq \alpha L_n((x-y)^2,y) + o(n^{-2})$$

Here, in the last inequality we used the definition of optimal. Since $\alpha < 0$, we find from Theorem 6.5 that

$$||f_n - L_n(f_n)||\ [-\delta_o,\delta_o] \geq |\alpha|\,||L_n(x-y)^2,y)||\ [x_n - C_1 n^{-1}\ ,\ x_n + C_1 n^{-1}] + o(n^{-2})$$

$$\geq |\alpha|\ C_2(1-x_n^2)n^{-2} + o(n^{-2}) \geq |\alpha|\ C_2(1-\delta_o^2)n^{-2} + o(n^{-2})$$

because the points x_n are in $[-\delta_o,\delta_o]$. This establishes (6.4.11) and the theorem is proved.

Corollary 6.3. If (L_n) is a sequence of positive polynomial operators which is optimal on $[-1,1]$, then (L_n) is saturated with order (n^{-2}).

Proof. Since (L_n) is optimal on $[-1,1]$, it is optimal on $[-\delta,\delta]$ for each $\delta < 1$. If $||f - L_n(f)||\ [-1,1] = o(n^{-2})$, then Theorem 6.6 shows that f is linear. Every function whose first derivative is in Lip 1 is approximated with order $O(n^{-2})$.

6.5. Notes.

6.5.1 The idea of solving certain extremal problems as a means of generating positive operators which will yield Jackson type theorems can be found in D.J. Newman and H.S. Shapiro$_1$ (see also DeVore$_1$ and Bojanic and DeVore$_1$). The extremal problem considered in Newman and Shapiro is different than that given in Theorem 6.2., but its solution also depends on the use of quadrature formulae. They generate operators which provide Jackson type estimates for approximation even in higher dimensions. The use of quadrature formulae to solve extremal problems is well established (see for example Karlin and Studden$_1$, pg. 312).

6.5.2 All the examples of optimal sequences given in this chapter were optimal from $C[-\frac{1}{2}, \frac{1}{2}]$ into $C[-\delta,\delta]$ for each $\delta < \frac{1}{2}$. Because they are in the form of convolution with an algebraic polynomial, they cannot be optimal from $C[-\frac{1}{2}, \frac{1}{2}]$ to $C[-\frac{1}{2}, \frac{1}{2}]$ (see the remarks in Section 5.1). There are no known examples of optimal sequences from $C[-\frac{1}{2}, \frac{1}{2}]$ to $C[-\frac{1}{2}, \frac{1}{2}]$. A natural form for such operators would be the interpolation type operators (see Section 5.1).

6.5.3 Corollary 6.2, when compared with Theorem 4.5 and Theorem 6.4 suggest that all sequences (L_n) which are optimal from $C[-1,1]$ to $C[-1,1]$ or more generally from $C[-1,1]$ to $C[-\delta,\delta]$, for each $\delta < 1$ should have as their saturation class $S(L_n) = \{f : f' \in \mathrm{Lip}\,1 \text{ on } [-1,1]\}$.

6.5.4 We have not been able to determine the saturation properties of the sequence of Hermite-Fejer operator, (H_n). There are at least two difficulties that arise in this problem. First, it appears from the estimates in Section 2.8 that the saturation order for (H_n) should include the factor $|T_n(x)|$, which vanishes on $(-1,1)$. Although this should not effect the determining of saturation order for (H_n) and the proving of a "o" theorem, this could effect the "O" theorem. Recall that all our theorems for characterizing the saturation class required that the function $\phi_n(x)$ in the saturation order (ϕ_n) should not vanish on (a,b).

The reason the "o" theorem is not yet known is that we do not have an asymptotic formula for (H_n). It is almost certain from the form of our estimates in Section 2.8 that an asymptotic formula for (H_n) would involve more than just a second order differential operator.

CHAPTER 7

APPROXIMATION OF CLASSES OF FUNCTIONS

7.1. Introduction.

Saturation is a way of measuring the efficiency of approximation by a sequence of operators (L_n). When (L_n) is saturated with order (ϕ_n) (or $(\phi_n(x))$), then we know that no amount of smoothness on the function f to be approximated can guarantee an error of approximation $o(\phi_n)$. However, saturation gives no information as to what functions are approximated with an order (Ψ_n), which is not optimal (i.e. $\Psi_n \neq O(\phi_n)$). There are two common ways of measuring the efficiency of approximation in the non optimal case: approximation of classes of functions, and inverse theorems. In this chapter, we will discuss problems in the approximation of classes of functions while in Chapter 8 we will deal with inverse theorems.

If A is a class of functions from $C^*[-\pi,\pi]$ (or $C[a,b]$), we define the error in approximating A by L_n to be

$$E(A,L_n) = \sup_{f\in A} ||f - L_n(f)|| \tag{7.1.1}$$

where of course the norm is taken in the range of L_n. If A is compact, then $E(A,L_n) \to 0$. In the case that, $L_n(1) = 1$, it is not necessary to have A compact for $E(A,L_n) \to 0$.

The problem in approximating classes of functions is to determine $E(A,L_n)$. For some operators L_n and classes A, this is possible, but generally this is too difficult. Instead, we will usually have to settle for an asymptotic analysis of $(E(A,L_n))$ as $n \to \infty$. For us, this analysis will take one of two forms. One is to find a sequence (Ψ_n) such that

$$(E(A,L_n)) \sim (\Psi_n) \qquad (7.1.2)$$

(See Section 3.1 for the notation of $\sim$). This, we call a weak asymptotic analysis of $E(A,L_n)$. A second more precise analysis is to find (Ψ_n) such that

$$E(A,L_n) = \Psi_n + o(\Psi_n) \ , \ (n \to \infty) \qquad (7.1.3).$$

We will call (7.1.3) a strong asymptotic analysis of $E(A,L_n)$.

It is clear that this is not the strongest possible asymptotic analysis of $E(A,L_n)$. However, finer estimates are usually more delicate and have only been obtained for specific examples.

If (L_n) is any sequence of positive operators on $C[a,b]$ and

$$\mu_n^2 = ||L_n((t-x)^2 \ , x)||$$

then for any class A which contains multiples of $1,x$ and x^2 , we know

$$E(A,L_n) \geq \mu_n^2 \qquad (7.1.4).$$

Also, if A is contained in the class $W^{(1)}(1,M) = \{f' \in Lip(1,M)\}$ then the estimate

$$E(A,L_n) \leq C \mu_n^2 \qquad (7.1.5)$$

follows from our estimates in Chapter 2 (Theorem 2.3). Thus, for example, the error in approximating any of the classes $A = W^{(r)}(\alpha,M)$ with $r \geq 2$ is

$$(E(A,L_n)) \sim (\mu_n^2) \quad .$$

Because of this, we will only consider smoothness up to the second derivative. Basically, we will concern ourselves with the classes $\mathrm{Lip}(\alpha,M)$, $\mathrm{Lip}^*(\alpha,M)$ and $W^{(1)}(\alpha,M)$ (see Section 1.3 for notation).

When (L_n) is a sequence of polynomial operators, then we have at our disposal the general theorems on approximation of classes of functions (see Section 1.6). For example, if (L_n) is a sequence of trigonometric polynomial operators on $C^*[-\pi,\pi]$ then

$$E(\mathrm{Lip}^*(\alpha,1)\ ,\ L_n) \geq cn^{-\alpha}$$

because of Theorem 1.8. Therefore, if (L_n) is optimal (see Section 4.1), then $E_n(\mathrm{Lip}^*(\alpha,1),L_n) \sim (n^{-\alpha})$.

7.2. Approximation of classes of functions for convolution operators. We wish first to consider the case of convolution operators on $C^*[-\pi,\pi]$. So, we assume $L_n(f) = f*d\mu_n$ with $d\mu_n$ an even, non-negative, Borel measure with unit mass on $[-\pi,\pi)$. In this case, a detailed analysis of approximation of classes of functions, especially $\mathrm{Lip}^*(2,M)$, is given in E. Gorlich and E. Stark$_2$.

Our objective is to develop some of the more useful techniques for determining the error in approximating classes and to give some applications to the better known operators.

If there is an even function $g \in A$, with $g(0) = 0$, such that

$$||f(x+t) + f(x-t) - 2f(x)|| \leq 2g(t) \quad , \quad 0 \leq t \leq \pi$$

for each $f \in A$, then we shall say g is an extreme function for A . It is clear that if g exists, it is unique. The importance of determining g is readily seen in the following theorem.

Theorem 7.1. *If A has an extreme function g then for $n \in N$*

$$E(A,L_n) = \frac{1}{\pi}\int_{-\pi}^{\pi} g(t)\, d\mu_n(t) \qquad (7.2.1).$$

Proof. If $f \in A$, and $x \in [-\pi,\pi]$, then

$$|f(x) - L_n(f,x)| = \left|\frac{1}{2\pi}\int_{-\pi}^{\pi} (f(x+t) + f(x-t) - 2f(x))d\mu_n(t)\right|$$

$$\leq \frac{1}{\pi}\int_{-\pi}^{\pi} g(t)\, d\mu_n(t)$$

and so

$$E(A,L_n) \leq \frac{1}{\pi}\int_{-\pi}^{\pi} g(t)\, d\mu_n(t) \qquad (7.2.2).$$

Now, $g \in A$ and

$$L_n(g,0) - g(0) = \frac{1}{\pi}\int_{-\pi}^{\pi} g(t)\, d\mu_n(t) \quad .$$

This shows that equality holds in (7.2.2).

For the classes $Lip^*(\alpha,1)$ and $Lip(\alpha,1)$, $0 < \alpha \leq 1$, it is easy to check that the function $g_\alpha(t) = |t|^\alpha$ is an extreme function. Thus, we have the following corollary.

Corollary 7.1. *If $0 < \alpha \leq 1$ then*

$$E(Lip(\alpha,1)\ ,\ L_n) = E(Lip^*(\alpha,1)\ ,\ L_n) = \frac{1}{\pi}\int_{-\pi}^{\pi} |t|^\alpha\, d\mu_n(t) \quad .$$

For $1 < \alpha \leq 2$, the situation is more subtle and some authors have made the mistake of saying $|t|^\alpha$ is an extreme function for $\mathrm{Lip}^*(\alpha,1)$ (see remarks by Gorlich and Stark$_2$, pg. 25). This is not correct since the periodic extension of $|t|^\alpha$ is not in $\mathrm{Lip}^*(\alpha,1)$ when $1 < \alpha \leq 2$. The following theorem determines the extreme functions for the classes $W^{(1)}(\alpha-1\ ,M)$, $1 < \alpha \leq 2$. Since $W^{(1)}(1,2) = \mathrm{Lip}^*(2,1)$, we determine the extreme function for $\mathrm{Lip}^*(2,M)$. For $1 < \alpha < 2$, the situation concerning the extreme functions for $\mathrm{Lip}^*(\alpha,M)$ is not settled.

Theorem 7.2. *If* $1 < \alpha \leq 2$, *then the function* g_α *defined by*

$$g_\alpha(t) = \begin{cases} |t|^\alpha & |t| \leq \frac{\pi}{2} \\ 2(\frac{\pi}{2})^\alpha - (\pi-|t|)^\alpha & \frac{\pi}{2} \leq |t| \leq \pi \end{cases} \tag{7.2.3}$$

is an extreme function for $W^{(1)}(\alpha-1\ ,\ \alpha 2^{2-\alpha})$.

Proof. Let $f \in W^{(1)}(\alpha-1,\alpha 2^{2-\alpha})$. If $0 \leq t \leq \frac{\pi}{2}$ and $x \in [-\pi,\pi]$, then

$$|f(x+t) + f(x-t) - 2f(x)| = |\int_0^t [f'(x+u) - f'(x-u)]du|$$

$$\leq \alpha 2^{2-\alpha} \int_0^t (2u)^{\alpha-1}\, du$$

$$= 2t^\alpha = 2g_\alpha(t) \quad .$$

Now, suppose for some $f \in W^{(1)}(\alpha-1,\alpha 2^{2-\alpha})$, we have

$$|f(x_o + t_o) + f(x_o - t_o) - 2f(x_o)| > 2\, g_\alpha(t_o) \tag{7.2.4}$$

with $x_o \in [-\pi,\pi]$ and $\frac{\pi}{2} < t_o \leq \pi$. We work for a contradiction. We suppose the term in the absolute value sign in (7.2.4) is positive. The other case is handled similarly or by replacing f by $-f$. Let

$$h(t) = \frac{1}{2}\,(f(x_o+t) + f(x_o-t) - 2f(x_o)) \quad .$$

It is easy to check that h is also in $W^{(1)}(\alpha-1,\alpha 2^{2-\alpha})$. Since $h(t_o) > g_\alpha(t_o)$ and $h(\frac{\pi}{2}) \leq g_\alpha(\frac{\pi}{2})$, we must have for some $\frac{\pi}{2} < \xi < t_o$

$$g_\alpha'(\xi) < h'(\xi) \quad .$$

Also $g_\alpha'(2\pi-\xi) = -g_\alpha'(\xi)$ and $h'(2\pi-\xi) = -h'(\xi)$, since g_α and h are both even and 2π periodic. Thus,

$$h'(2\pi-\xi) - h'(\xi) < g_\alpha'(2\pi-\xi) - g_\alpha'(\xi) = -2g_\alpha'(\xi)$$

$$= -2\alpha(\pi-\xi)^{\alpha-1} = -\alpha 2^{2-\alpha}\,(2\pi-2\xi)^{\alpha-1}$$

which contradicts the fact that h is in $W^{(1)}(\alpha-1,\alpha 2^{2-\alpha})$, as desired. It is easy to see that g_α is indeed in the class $W^{(1)}(\alpha-1,\alpha 2^{2-\alpha})$ which proves the theorem.

<u>Corollary 7.2</u>. <u>If</u> $1 < \alpha \leq 2$, <u>then</u> <u>for</u> $n \in N$

$$E(W^{(1)}(\alpha-1,\alpha 2^{2-\alpha}),L_n) = \frac{1}{\pi}\int_{-\pi}^{\pi} g_\alpha(t)\,d\mu_n(t)$$

<u>where</u> g_α <u>is</u> <u>defined</u> <u>by</u> (7.2.3).

It is easy to calculate the Fourier coefficients of $|t|$ and $g_2(t)$ and thereby obtain the following expansion for $E(A,L_n)$ when $A = \mathrm{Lip}^*(\alpha,1)$,

$\alpha = 1,2$.

Corollary 7.3. *If* $(\rho_{k,n})_{k=0}^{\infty}$ *are the real Fourier coefficients of* $d\mu_n$ *then*

$$E(\mathrm{Lip}^*(1,1),L_n) = \frac{1}{2}\pi - \frac{4}{\pi}\sum_{k=0}^{\infty}(2k+1)^{-2}\rho_{2k+1,n} \qquad (7.2.5)$$

$$E(\mathrm{Lip}^*(2,1),L_n) = \frac{\pi^2}{4} - \frac{8}{\pi}\sum_{k=0}^{\infty}(-1)^k(2k+1)^{-3}\rho_{2k+1,n} \qquad (7.2.6).$$

When $0 < \alpha < 1$, the Fourier coefficients of $|t|^{\alpha}$ can be determined asymptotically, but then it is usually better to try to asymptotically determine $\frac{1}{\pi}\int_{-\pi}^{\pi}|t|^{\alpha}\,d\mu_n(t)$ directly, as we will see in the examples of the next section.

Although we do not have an explicit determination of $E(\mathrm{Lip}^*(\alpha,1),L_n)$ in terms of extreme functions when $1 < \alpha < 2$, it is easy to see that $E(\mathrm{Lip}^*(\alpha,1),L_n)$ is weakly asymptotic to $\frac{1}{\pi}\int_{-\pi}^{\pi}|t|^{\alpha}\,d\mu_n(t)$.

Theorem 7.3. *If* $1 < \alpha \le 2$, *then for each* $n \in N$

$$\frac{2^{-\alpha}}{\pi}\int_{-\pi}^{\pi}|t|^{\alpha}\,d\mu_n(t) \le E(\mathrm{Lip}^*(\alpha,1),L_n) \le \frac{1}{\pi}\int_{-\pi}^{\pi}|t|^{\alpha}\,d\mu_n(t) \qquad (7.2.7).$$

If in addition, (L_n) *satisfies any of the equivalent conditions in Theorem* 3.8, *then*

$$E(\mathrm{Lip}^*(\alpha,1),L_n) = \frac{1}{\pi}\int_{-\pi}^{\pi}|t|^{\alpha}\,d\mu_n(t) + o\Big(\frac{1}{\pi}\int_{-\pi}^{\pi}|t|^{\alpha}\,d\mu_n(t)\Big) \qquad (7.2.8).$$

Proof. The right hand side of (7.2.7) follows directly from the definition of $\mathrm{Lip}^*(\alpha,1)$. To see the left hand side, we note first that the function $g_{\alpha}(t)$ given in (7.2.3) is in $\mathrm{Lip}^*(\alpha,1)$. Also, $2^{-\alpha}|t|^{\alpha} \le g_{\alpha}(t)$. Thus

$$\frac{2^{-\alpha}}{\pi} \int_{-\pi}^{\pi} |t|^\alpha \, d\mu_n(t) \le \frac{1}{\pi} \int_{-\pi}^{\pi} g_\alpha(t) \, d\mu_n(t)$$

$$= L_n(g_\alpha, 0) - g_\alpha(0)$$

$$\le E(\mathrm{Lip}^*(\alpha,1), L_n) \ .$$

To see (7.2.8), we need only recall that if (L_n) satisfies one of the conditions in Theorem 3.8, then

$$\int_{\pi/2}^{\pi} d\mu_n(t) = o\left(\int_{-\pi}^{\pi} t^2 \, d\mu_n(t)\right) = o\left(\int_{-\pi}^{\pi} |t|^\alpha \, d\mu_n(t)\right) \ .$$

Therefore,

$$E(\mathrm{Lip}^*(\alpha,1), L_n) \ge \frac{1}{\pi} \int_{-\pi}^{\pi} g_\alpha(t) \, d\mu_n(t) = \frac{1}{\pi} \int_{-\pi/2}^{\pi/2} |t|^\alpha \, d\mu_n(t) + o\left(\frac{1}{\pi} \int_{-\pi}^{\pi} |t|^\alpha \, d\mu_n(t)\right)$$

$$= \frac{1}{\pi} \int_{-\pi}^{\pi} |t|^\alpha \, d\mu_n(t) + o\left(\frac{1}{\pi} \int_{-\pi}^{\pi} |t|^\alpha d\mu_n(t)\right) \ .$$

7.3. Some examples. We have shown in the last section that the asymptotic behavior of $E(\mathrm{Lip}^*(\alpha,1), L_n)$ is linked to the integrals $\frac{1}{\pi} \int_{-\pi}^{\pi} |t|^\alpha \, d\mu_n(t)$. We will now give some examples, which will indicate what is generally involved in the calculation of these integrals.

First, let's consider the Fejer operators (σ_n) .

Theorem 7.4. *For the Fejer operators, we have*

$$E(\mathrm{Lip}^*(1,1), \sigma_n) = \frac{\pi}{2} - \frac{4}{\pi} \sum_{k=0}^{[\frac{1}{2} n]} (2k+1)^{-2} + \frac{4}{(n+1)\pi} \sum_{k=0}^{[\frac{1}{2} n]} (2k+1)^{-1}$$

$$= \frac{2 \ln n}{\pi n} + O(n^{-1}) \qquad (7.3.1)$$

and

$$E(\mathrm{Lip}^*(2,1),\sigma_n) = \left(\frac{\pi^2}{4} - \frac{8}{\pi}\sum_{k=0}^{[\frac{1}{2}n]} (-1)^k(2k+1)^{-3} + \frac{8}{(n+1)\pi}\sum_{k=0}^{[\frac{1}{2}n]} (-1)^k(2k+1)^{-2}\right.$$

$$= \frac{4}{(n+1)\pi}\sum_{k=0}^{\infty} (-1)^k(2k+1)^{-2} + O(n^{-2}) \qquad (7.3.2).$$

Proof. First, from (7.2.5) we find

$$E(\mathrm{Lip}^*(1,1),\sigma_n) = \frac{\pi}{2} - \frac{4}{\pi}\sum_{k=0}^{[\frac{1}{2}n]} \left(1 - \frac{2k+1}{n+1}\right)(2k+1)^{-2}$$

$$= \left(\frac{\pi}{2} - \frac{4}{\pi}\sum_{k=0}^{[\frac{1}{2}n]} (2k+1)^{-2}\right) + \frac{4}{(n+1)\pi}\sum_{k=0}^{[\frac{1}{2}n]} (2k+1)^{-1} \qquad (7.3.3)$$

where we have used the fact that the coefficients $\rho_{k,n}$ for σ_n are $1 - \frac{k}{n+1}$, $k \le n+1$ and 0 for $k > n+1$. Now, for any m

$$\sum_{k=0}^{m} (2k+1)^{-2} = \sum_{k=1}^{2m+1} k^{-2} - \sum_{k=1}^{m} (2k)^{-2} = \frac{\pi^2}{8} + O(m^{-1}) \qquad (7.3.4)$$

where we realize that $\sum_1^{\infty} k^{-2} = \frac{\pi^2}{6}$. This shows that the first term in (7.3.3) is $O(n^{-1})$. To estimate the second term, we write

$$\sum_{k=0}^{m} (2k+1)^{-1} = \sum_{k=1}^{2m+1} k^{-1} - \sum_{k=1}^{m} (2k)^{-1}$$

$$= \ln(2m+1) - \frac{1}{2}\ln(m) + O(1)$$

$$= \ln m + \ln 2 - \frac{1}{2}\ln m + O(1)$$

$$= \frac{1}{2}\ln m + O(1) \quad .$$

Thus, with $m = [\frac{1}{2} n]$

$$\sum_{k=0}^{[\frac{1}{2} n]} (2k+1)^{-1} = \frac{1}{2} \ln[\frac{1}{2} n] + O(1)$$

$$= \frac{1}{2} \ln n + O(1) \quad .$$

Putting this and (7.3.4) into (7.3.3) we obtain (7.3.1). To obtain (7.3.2), we use (7.2.6), so that

$$E(\mathrm{Lip}^*(2,1),\sigma_n) = \frac{\pi^2}{4} - \frac{8}{\pi} \sum_{k=0}^{[\frac{1}{2} n]} \frac{(-1)^k}{(2k+1)^3} \left(1 - \frac{2k+1}{n+1}\right)$$

$$= \left(\frac{\pi^2}{4} - \frac{8}{\pi} \sum_{k=0}^{[\frac{1}{2} n]} (-1)^k (2k+1)^{-3}\right)$$

$$+ \frac{8}{(n+1)\pi} \sum_{k=0}^{[\frac{1}{2} n]} (-1)^k (2k+1)^{-2} \qquad (7.3.5).$$

Since, $\frac{\pi^3}{32} = \sum_{k=1}^{\infty} (-1)^k (2k+1)^{-3}$, (see K. Knopp$_1$, pg. 240), the second part of (7.3.2) easily follows from (7.3.5).

For $0 < \alpha < 1$ and $1 < \alpha < 2$, we do not use a Fourier coefficient argument but rather estimate $\frac{1}{\pi} \int_{-\pi}^{\pi} |t|^{\alpha} F_n(t)\, dt$ directly where F_n is the Fejer kernel. We define $I(p,q)$ as

$$I(p,q) = \frac{1}{\pi} \int_0^{\infty} t^p \sin^q t \, dt \quad .$$

<u>Theorem 7.5.</u> <u>If</u> $0 < \alpha < 1$

$$E(\mathrm{Lip}^*(\alpha,1),\sigma_n) = \frac{2^{\alpha+1}}{(n+1)^{\alpha}} I(\alpha-2,2) + O(n^{-1}) \qquad (7.3.6).$$

If $1 < \alpha < 2$,

$$E(\mathrm{Lip}^*(\alpha,1),\sigma_n) \sim n^{-1} \qquad (7.3.7).$$

Proof. First, for $0 < \alpha < 1$, we have from Corollary 7.1 that

$$E(\mathrm{Lip}^*(\alpha,1),\sigma_n) = \frac{1}{(n+1)\pi} \int_0^\pi t^\alpha \left(\frac{\sin(\frac{n+1}{2})\ t}{\sin \frac{t}{2}} \right)^2 dt \quad .$$

Since $|4t^{-2} - \sin^{-2} \frac{t}{2}|$ is bounded on $[0,\pi]$, we have

$$E(\mathrm{Lip}^*(\alpha,1),\sigma_n) = \frac{1}{(n+1)\pi} \int_0^\pi 4t^{\alpha-2} \sin^2(\tfrac{n+1}{2})t\ dt + O(\tfrac{1}{n\pi} \int_0^\pi t^\alpha \sin^2(\tfrac{n+1}{2})t\, dt)$$

$$= \frac{4}{(n+1)\pi} \int_0^\pi t^{\alpha-2} \sin^2(\tfrac{n+1}{2})t\ dt + O(n^{-1}) \quad .$$

By making the change of variable $(\frac{n+1}{2})t = u$, we find

$$E(\mathrm{Lip}^*(\alpha,1),\sigma_n) = \frac{2^{\alpha+1}}{\pi(n+1)^\alpha} \int_0^{(\frac{n+1}{2}\pi)} u^{\alpha-2} \sin^2 u\ du + O(n^{-1}) \quad .$$

The estimate (7.3.6) then follows from the fact that

$$(n+1)^{-\alpha} \int_{(\frac{n+1}{2}\pi)}^\infty u^{\alpha-2} \sin^2 u\ du \le (n+1)^{-\alpha} \int_{(\frac{n+1}{2}\pi)}^\infty u^{\alpha-2} = O(n^{-1}) \ .$$

Now, we turn to the estimate (7.3.7). Here the integrals

$$\frac{2}{\pi} \int_0^\pi t^\alpha F_n(t)\ dt$$

only provide a weak asymptotic estimate of $E(\mathrm{Lip}^*(\alpha,1),\sigma_n)$, via Theorem 7.3. We have for $1 < \alpha < 2$,

$$\frac{1}{\pi(n+1)} \int_0^\pi t^\alpha \left(\frac{\sin \frac{(n+1)}{2} t}{\sin \frac{t}{2}} \right)^2 dt \sim (n+1)^{-1} \int_0^\pi t^{\alpha-2} \sin^2(\frac{n+1}{2} t)\, dt \quad .$$

This gives

$$E(Lip^*(\alpha,1),\sigma_n) \sim (n+1)^{-1} \int_0^\pi t^{\alpha-2} \sin^2(\frac{n+1}{2} t)\, dt$$

$$= \frac{1}{2}(n+1)^{-1} \int_0^\pi t^{\alpha-2}(1 - \cos(n+1)t) = \frac{1}{2}(n+1)^{-1} \int_0^\pi t^{\alpha-2} dt + o(n^{-1})$$

which gives (7.3.7). Note that the Fourier coefficients of $t^{\alpha-2}$ tend to 0.

We should note that it follows directly from the general theorems on approximation of classes of functions by trigonometric polynomials, (Theorem 1.8) that for some $C > 0$, $E(Lip^*(\alpha,1),\sigma_n) \geq Cn^{-\alpha}$, when $0 < \alpha < 1$, $n \in N$.

We will now give similar estimates for the Jackson-Matsuoaka operators (see Section 3.10).

<u>Theorem 7.6.</u> <u>If</u> $p \geq q \geq 2$, <u>then for the Jackson-Matsuoaka operators</u> $J_{n,p,q}$, <u>we have for</u> $0 < \alpha \leq 2$

$$E(Lip^*(\alpha,1),J_{n,p,q}) = 2^\alpha \frac{I(\alpha-2q,2p)}{I(2q,2p)} (n+1)^{-\alpha} + o(n^{-\alpha}) \quad (n\to\infty) \qquad (7.3.8).$$

<u>Proof</u>. Here, we know the kernels $K_{n,p,q}$ satisfy the equivalent conditions in Theorem 3.8 (see Section 3.10), so that when $1 < \alpha \leq 2$, we can use (7.2.8) to estimate $E(Lip^*(\alpha,1),J_{n,p,q})$. In the following estimates, we take $0 \leq \alpha \leq 2$.

Since $|(2t^{-1})^{2q} - \sin^{-2q} \frac{t}{2}|$ is bounded on $[o,\pi]$, we find

$$\frac{2}{\pi}\int_0^{\pi} t^{\alpha} K_{n,p,q}(t)dt = \frac{2}{\pi} C_{n,p,q} \int_0^{\pi} t^{\alpha} \frac{\sin^{2p}(\frac{n+1}{2}t)}{\sin^{2q}\frac{t}{2}} dt$$

$$= \frac{2^{2q+1}}{\pi} C_{n,p,q} \int_0^{\pi} t^{\alpha-2q} \sin^{2p}(\frac{n+1}{2}t)dt + O(C_{n,p,q})$$

$$= \frac{2^{2q+1}}{\pi} C_{n,p,q} \int_0^{\pi} t^{\alpha-2q} \sin^{2p}(\frac{n+1}{2}t)dt + o(n^{-2}) \quad (7.3.9).$$

Here, we have used the fact that $C_{n,p,q} = o(n^{-2})$ which was given in Section 3.10.

Changing variables in the integral by letting $u = (\frac{n+1}{2}t)$, we find

$$\frac{2}{\pi}\int_0^{\pi} t^{\alpha} K_{n,p,q}(t)dt = \frac{2^{2q+1}}{\pi}(\frac{2}{n+1})^{\alpha-2q+1} C_{n,p,q} \int_0^{(\frac{n+1}{2}\pi)} u^{\alpha-2q} \sin^{2p}u\, du + o(n^{-2})$$

$$= 2^{\alpha+2} C_{n,p,q}(n+1)^{2q-1-\alpha} I(\alpha-2q,2p) + O(C_{n,p,q}) + o(n^{-2})$$

$$= 2^{\alpha+2} C_{n,p,q} (n+1)^{2q-1-\alpha} I(\alpha-2q,2p) + o(n^{-2}) \quad (7.3.10)$$

where we have estimated the remainder using the fact that

$$\int_{(\frac{n+1}{2}\pi)}^{\infty} u^{\alpha-2q} \sin^{2p} u\, du \le \int_{(\frac{n+1}{2}\pi)}^{\infty} u^{\alpha-2q} = O(n^{\alpha-2q+1}) .$$

In particular, when $\alpha = 0$, we find

$$1 = 4\, C_{n,p,q}(n+1)^{2q-1}\; I(-2q,2p) + o(n^{-2})$$

which shows that $C_{n,p,q} = \frac{1}{4}\frac{(n+1)^{-2q+1}}{I(-2q,2p)} + o(n^{-2q-1})$. Using this in (7.3.10), we find for $0 < \alpha \leq 2$

$$\frac{2}{\pi}\int_0^\pi t^\alpha K_{n,p,q}(t)\,dt = 2^\alpha(n+1)^{-\alpha}\frac{I(\alpha-2q,2p)}{I(-2q,2p)} + o(n^{-2}) \quad .$$

Finally, using Corollary 7.1 when $0<\alpha\leq 1$ and (7.2.8) when $1<\alpha\leq 2$, we find

$$E(Lip^*(\alpha,1),J_{n,p,q}) = \frac{2}{\pi}\int_0^\pi t^\alpha K_{n,p,q}(t)\,dt + o(n^{-\alpha})$$

$$= 2^\alpha\,(n+1)^{-\alpha}\frac{I(\alpha-2q,2p)}{I(-2q,2p)} + o(n^{-\alpha}) \quad .$$

<u>7.4. Approximation of classes for</u> $C[a,b]$. Suppose (L_n) is a sequence of positive operators from $C[a,b]$ into $C[c,d]$ with $[c,d] \subseteq [a,b]$. Generally, we will not have $L_n(e_o) = e_o$. This will make it necessary to require that our class of functions is compact. Therefore, we will work with the classes $Lip(\alpha,M,M_o)$ (see Section 1.3) in place of $Lip(\alpha,M)$.

The following theorem for determining $E(Lip(\alpha,M,M_o),L_n)$ is the analogue of Theorem 7.3.

<u>Theorem 7.7.</u> *Let (L_n) be a sequence of positive linear operators from $C[a,b]$ into $C[c,d]$ with $[c,d] \subseteq [a,b]$ such that*

$$||e_i - L_n(e_i)|| = o(\mu_n^2) \quad , \quad i = 0,1 \tag{7.4.1}$$

where

$$\mu_n^2 = ||L_n((t-x)^2,x)|| \tag{7.4.2}$$

If $0 < \alpha \leq 1$, *and* $M_o \geq |b-a|^\alpha$ *then*

$$E(\mathrm{Lip}(\alpha,1,M_o),L_n) = ||L_n(|t-x|^\alpha,x)|| + o(\mu_n^2) \qquad (7.4.3).$$

<u>If</u> $1 < \alpha \leq 2$, <u>and</u> $M_o \geq |b-a|^\alpha$ <u>then</u>

$$E(W^{(1)}(\alpha-1,\alpha,M_o),L_n) \sim ||L_n(|t-x|^\alpha,x)|| \qquad (7.4.4).$$

<u>Remarks</u>. In the event that $L_n(e_o) = e_o$ then the o term in (7.4.3) can be dropped. The classes are defined over $[a,b]$. Observe that $\mu_n^2 = O(||L_n(|t-x|^\alpha,x)||)$.

<u>Proof</u>. First, we consider (7.4.3). If $f \in \mathrm{Lip}(\alpha,1,M_o)$ then

$$|f(t) - f(x)| \leq |t-x|^\alpha \quad t \in [a,b] \quad , \quad x \in [c,d] \quad .$$

Therefore,

$$|L_n(f,x) - f(x)| \leq |L_n(f(t) - f(x),x)| + |L_n(f(x),x) - f(x)|$$

$$\leq L_n(|t-x|^\alpha,x) + M_o \; ||L_n(e_o) - e_o||$$

$$\leq ||L_n(|t-x|^\alpha,x)|| + o(\mu_n^2)$$

where for the second term we have used (7.4.1). Taking a supremum over $x \in [c,d]$, and using the fact that f was arbitrary, we find

$$E(\mathrm{Lip}(\alpha,1,M_o),L_n) \leq ||L_n(|t-x|^\alpha,x)|| + o(\mu_n^2) \quad .$$

For each $x_o \in [c,d]$, $|t-x_o|^\alpha$ is in $\mathrm{Lip}(\alpha,1,M_o)$. Here is where we use the restriction $M_o \geq |b-a|^\alpha$. Thus

$$L_n(|t-x_o|^\alpha,x_o) \le E(\mathrm{Lip}(\alpha,1,M_o),L_n)$$

and so

$$||L_n(|t-x|^\alpha,x)|| \le E(\mathrm{Lip}(\alpha,1,M_o),L_n) \quad .$$

We have now established (7.4.3).

For (7.4.4), we argue similarly. Let $f \in W^{(1)}(\alpha-1,\alpha,M_o)$. If $x \in [c,d]$ and $t \in [a,b]$ with $t > x$, then

$$|f(t) - f(x) - f'(x)(t-x)| \le |\int_x^t (f'(u) - f'(x))\,du| \le \alpha \int_x^t |u-x|^{\alpha-1}du$$
$$\le |t-x|^\alpha$$

The same estimate holds when $t < x$. Therefore,

$$|L_n(f,x) - f(x)| \le |L_n(f(t) - f(x) - f'(x)(t-x),x)|$$
$$+ |L_n(f(x),x) - f(x)| + |L_n(f'(x)(t-x),x)|$$
$$\le ||L_n(|t-x|^\alpha,x)|| + M_o||e_o - L_n(e_o)||$$
$$+ |f'(x)|\ ||e_1 - L_n(e_1)||$$
$$\le ||L_n(|t-x|^\alpha,x)|| + o(\mu_n^2) \quad . \qquad (7.4.5).$$

For the last two terms we have used (7.4.1). Also, we used the fact that $||f'|| \le M'$, where M' does not depend on f . This is established from the identity

$$|f'(x)(t-x)| = |f(t) - f(x) - \int_x^t (f'(u) - f'(x))du| \le 2M_0 + |t-x|^\alpha .$$

Given x , we choose t so that $|x-t| = \frac{b-a}{2}$. Then

$$|f'(x)(\tfrac{b-a}{2})| \le 2M_0 + |\tfrac{b-a}{2}|^\alpha$$

or

$$||f'|| \le 4M_0(b-a)^{-1} + |\tfrac{b-a}{2}|^{\alpha-1} = M' .$$

Taking a supremum over $x \in [c,d]$ on the left hand side of (7.4.5) will yield

$$||L_n(f) - f|| \le ||L_n(|t-x|^\alpha,x) + o(u_n^2) .$$

Therefore,

$$E(W^{(1)}(\alpha-1,\alpha,M_0),L_n) \le ||L_n(|t-x|^\alpha,x)|| + o(u_n^2) \qquad (7.4.6).$$

The functions $2^{\alpha-2}|t-x_0|^\alpha$ are in $W^{(1)}(\alpha-1,\alpha,M_0)$ whenever $x_0 \in [c,d]$, so that

$$2^{\alpha-2}||L_n(|t-x|^\alpha,x)|| \le E(W^{(1)}(\alpha-1,\alpha,M_0),L_n) .$$

This coupled with (7.4.6) and $\mu_n^2 = O(||L_n(|t-x|^\alpha,x)||)$ shows (7.4.4).

For an application of Theorem 7.7, consider the Landau operators ℓ_n introduced in Section 2.5.

<u>Theorem 7.8.</u> <u>Let</u> $0 < \delta < \frac{1}{2}$, <u>so that</u> ℓ_n <u>maps</u> $C[-\frac{1}{2}, \frac{1}{2}]$ <u>into</u> $C[-\delta,\delta]$.

Then, if $0 < \alpha \leq 1$ and $M_o \geq 1$

$$E(\mathrm{Lip}(\alpha,1,M_o),\ell_n) = \frac{\Gamma(\frac{\alpha+1}{2})\quad\Gamma(n+\frac{3}{2})}{\Gamma(\frac{1}{2})\Gamma(n+\frac{3}{2}+\frac{1}{2}\alpha)} + o(n^{-1}) \sim n^{-\frac{\alpha}{2}}$$

and if $1 < \alpha < 2$, $M_o \geq 1$

$$E(W^{(1)}(\alpha-1,\alpha,M_o),\ell_n) \sim \frac{\Gamma(\frac{\alpha+1}{2})\quad\Gamma(n+\frac{3}{2})}{\Gamma(\frac{1}{2})\ \Gamma(n+\frac{3}{2}+\frac{\alpha}{2})} \sim n^{-\frac{\alpha}{2}}$$

Proof. We use Theorem 7.7. If $x \in [-\delta,\delta]$, and $0 < \alpha \leq 2$, then

$$\begin{aligned}\ell_n(|t-x|^\alpha,x) &= c_n \int_{-1/2}^{1/2} |t-x|^\alpha (1-(t-x)^2)^n\, dt \\ &= c_n \int_{-1/2-x}^{1/2-x} |t|^\alpha (1-t^2)^n\, dt \qquad (7.4.7).\end{aligned}$$

We have seen in (2.5.3) that

$$c_n \int_{-1}^{1} t^4(1-t^2)^n\, dt = o(n^{-1})$$

from which it follows

$$c_n \int_{1/2-\delta}^{1} (1-t^2)^n\, dt \leq (\frac{1}{2}-\delta)^{-4}\, c_n \int_{-1}^{1} t^4(1-t^2)^n\, dt = o(n^{-1}) \quad .$$

Therefore, (7.4.7) can be written as

$$\begin{aligned}\ell_n(|t-x|^\alpha,x) &= c_n \int_{-1/2+\delta}^{1/2-\delta} |t|^\alpha (1-t^2)^n\, dt + o(n^{-1}) \\ &= c_n \int_{-1}^{1} |t|^\alpha (1-t^2)^n\, dt + o(n^{-1})\end{aligned}$$

$$= c_n \int_0^1 u^{\frac{1}{2}\alpha - \frac{1}{2}} (1-u)^n \, du + o(n^{-1})$$

$$= \frac{\Gamma(\frac{\alpha+1}{2})\,\Gamma(n+1)}{\Gamma(n + \frac{3}{2} + \frac{\alpha}{2})} \, c_n + o(n^{-1})$$

with the "o" term uniform in $x \in [-\delta,\delta]$. We use the value of c_n which is given in (2.5.3) to find

$$\ell_n(|t-x|^{\alpha},x) = \frac{\Gamma(\frac{\alpha+1}{2})\,\Gamma(n + \frac{3}{2})}{\Gamma(\frac{1}{2})\,\Gamma(n + \frac{3}{2} + \frac{\alpha}{2})} + o(n^{-1}) \quad .$$

Taking a supremum over $x \in [-\delta,\delta]$ and invoking Theorem 7.7 gives the desired result.

As another example, we consider the Bernstein operators. We first want to show that for a suitably small constant $0 < a < \frac{1}{4}$, we have for n sufficiently large

$$\sum_{k \in I_n} p_{n,k}(\tfrac{1}{2}) \le \tfrac{1}{2} \tag{7.4.8}$$

where $I_n = \{k : |\frac{k}{n} - \frac{1}{2}| \le a\, n^{-1/2}\}$. To establish (7.4.8) we will use the following two inequalities:

<u>7.4.9</u> There are constants $C_1, C_2 > 0$ such that for $n \in N$

$$C_1 n^{n + \frac{1}{2}} e^{-n} \le n! \le C_2 n^{n + \frac{1}{2}} e^{-n}$$

<u>7.4.10</u> There are constants C_3 and $C_4 > 0$ such that for $n \ge 2$

$$C_3 \le (1 + n^{-1}x)^n \le C_4 \quad , \quad x \in [-1,1] \quad .$$

The first of these inequalities follows from Stirling's formula and the second from the uniform convergence of $(1 + n^{-1}x)^n$ to e^x on $[-1,1]$.

It will be simpler in what follows, if we let C always denote a positive constant which is independent of any of the parameters involved. Returning to (7.4.8), we first find that

$$\sum_{k\varepsilon I_n} p_{n,k}(\tfrac{1}{2}) \leq (2an^{1/2} + 1) \max_{k\varepsilon I_n} p_{n,k}(\tfrac{1}{2}) \tag{7.4.11}$$

since the number of indices k in I_n is $\leq (2an^{1/2} + 1)$. Now, from (7.4.9), it follows that

$$p_{n,k}(\tfrac{1}{2}) = (\tfrac{1}{2})^n \binom{n}{k} \leq C(\tfrac{1}{2})^n \left(\frac{n}{k(n-k)}\right)^{1/2} \left(\frac{n}{k}\right)^k \left(\frac{n}{n-k}\right)^{n-k} \tag{7.4.12}$$

To estimate $\left(\frac{n}{k(n-k)}\right)^{1/2}$, we need only observe our restriction on k to see that

$$\left(\frac{n}{k(n-k)}\right)^{1/2} \leq n^{1/2} \left(\tfrac{1}{2} n - an^{1/2}\right)^{-1} \leq Cn^{-1/2} \tag{7.4.13}$$

Recall that we only consider $a < \frac{1}{4}$.

In view of these last three inequalities, we see that it is sufficient to show that for $k \varepsilon I_n$

$$\left(\frac{n}{k}\right)^k \left(\frac{n}{n-k}\right)^{n-k} \leq C\, 2^n \tag{7.4.14}$$

Since then taking a suitably small in (7.4.11) will realize (7.4.8) for n sufficiently large. The inequality (7.4.14) is proved by writing $k = \frac{1}{2} n + \ell$, where $-an^{1/2} \leq \ell \leq an^{1/2}$ (ℓ need not be an integer) and finding

$$\left(\frac{n}{k}\right)^k\left(\frac{n}{n-k}\right)^{n-k} = 2^n(1 + 2\ell n^{-1})^{-\frac{1}{2}n-\ell}(1-2\ell n^{-1})^{-\frac{1}{2}n+\ell}$$

$$= 2^n(1 + 2\ell n^{-1})^{-\frac{1}{2}n}(1 - 2\ell n^{-1})^{-\frac{1}{2}n} \times$$

$$\times (1 - 2\ell n^{-1})^{\ell}(1 + 2\ell n^{-1})^{-\ell} \qquad (7.4.15).$$

From (7.4.10) we find

$$(1 + 2\ell n^{-1})^{-\frac{1}{2}n}(1 - 2\ell n^{-1})^{-\frac{1}{2}n} = (1 - 4\ell^2 n^{-2})^{-\frac{1}{2}n} \leq C .$$

Here, we have used the fact that $\ell \leq an^{1/2} \leq \frac{n^{1/2}}{4}$. Also, from (7.4.10), it follows that

$$(1 - 2\ell n^{-1})^{\ell}(1 + 2\ell n^{-1})^{-\ell} = \left(1 - \frac{2\ell^2 n^{-1}}{\ell}\right)^{\ell}\left(1 + \frac{2\ell^2 n^{-1}}{\ell}\right)^{\ell} \leq C .$$

Using these last two inequalities in (7.4.15) gives (7.4.14), which was desired.

Now, it is easy to give weak asymptotic estimates for the error in approximating classes by Bernstein polynomials.

Theorem 7.9. *If* $0 < \alpha \leq 1$, *then considering* $B_n : C[0,1] \to C[0,1]$, *we have*

$$E(\mathrm{Lip}(\alpha,1,M_o),B_n) \sim n^{-1/2\,\alpha}$$

and if $1 < \alpha \leq 2$, *then*

$$E(W^{(1)}(\alpha-1,\alpha,M_o),B_n) \sim n^{-1/2\,\alpha}$$

whenever $M_o \geq 1$.

<u>Proof</u>. By virtue of Theorem 7.7, we need only show that if $0 < \alpha \leq 2$, then

$$||B_n(|t-x|^\alpha,x)|| \sim n^{-1/2\,\alpha} .$$

The estimate

$$||B_n(|t-x|^\alpha,x)|| \leq Cn^{-1/2\,\alpha}$$

follows from the standard estimates (2.4.3) and (2.4.4). A lower estimate follows readily from (7.4.8). If $0 < a$ is chosen so that (7.4.8) holds for n sufficiently large then

$$B_n(|t-\tfrac{1}{2}|^\alpha,\tfrac{1}{2}) = \sum_{k=0}^{n} |\tfrac{k}{n}-\tfrac{1}{2}|^\alpha \, p_{n,k}(\tfrac{1}{2}) \geq (an^{-1/2})^\alpha \sum_{k\notin I_n} p_{n,k}(\tfrac{1}{2})$$

$$\geq \tfrac{1}{2}\,(an^{-1/2})^\alpha$$

where we have used the fact that

$$\sum_{k=0}^{n} p_{n,k}(\tfrac{1}{2}) = 1 .$$

<u>7.5. Connections between the degree of approximation and the modulus of continuity</u>. Our fundamental estimates developed in Chapter 2 generally took the form

$$||f-L_n(f)|| \leq C\omega(f,\mu_n)$$

where in the periodic case

$$\mu_n^2 = ||L_n(\sin^2(\tfrac{t-x}{2}),x)||$$

and in the non-periodic case

$$\mu_n^2 = ||L_n((t-x)^2,x)|| \quad .$$

By now, it is clear that these estimates are not always the best possible. Improvements have been given in this chapter. For example, for the Fejér operators (σ_n), $2\mu_n^2 = n^{-1}$ but from Theorem 7.5

$$E(\mathrm{Lip}^*(\alpha,M),\sigma_n) \sim \begin{cases} n^{-\alpha} & \text{if } 0 < \alpha < 1 \\ n^{-1}\ln n & \alpha = 1 \\ n^{-1} & 1 < \alpha \leq 2 \end{cases} \qquad (7.5.1).$$

An important observation is that it is not possible to put (7.5.1) in the form

$$E(\mathrm{Lip}^*(\alpha,M),\sigma_n) \sim \phi_n^\alpha$$

with (ϕ_n) a fixed sequence not dependent on α. In this sense, it is not completely clear how the error in approximating a class of functions by (σ_n) is related to the smoothness condition which determines the class. It is sometimes possible to give finer estimates which reflect this dependence. We will illustrate this in two cases; the first of which is the following result of S.B. Stečkin[1] for the Fejér operators.

<u>Theorem 7.10. There are constants</u> $C_1, C_2 > 0$ <u>such that whenever</u> ω <u>is a modulus of continuity on</u> $[0,\pi]$

$$C_1 M n^{-1} \sum_{k=1}^{n} \omega(k^{-1}) \leq E(C_\omega(M),\sigma_n) \leq C_2 M n^{-1} \sum_{k=1}^{n} \omega(k^{-1}) \qquad (7.5.2).$$

<u>Remark.</u> Here, $C_\omega(M)$ denotes the space of all functions $f \in C^*$ for which

$$||f(x+t) - f(x)|| \le M\,\omega(t) \ , \quad 0 < t \le \pi \ .$$

We will incorporate part of the proof in the following lemma which we will also use in the second example.

<u>Lemma 7.1.</u> <u>Let</u> ω <u>be a modulus of continuity defined on</u> $[0,\pi]$. <u>If</u> $n \ge 2$ <u>then</u>

$$\pi n^{-1} \int_{\pi n^{-1}}^{\pi} t^{-2}\,\omega(t)\,dt \le \sum_{k=1}^{n-1} k^{-2}\,\omega(n^{-1}(k+1)\pi)$$

$$\le 8\pi n^{-1} \int_{\pi n^{-1}}^{\pi} t^{-2}\,\omega(t)\,dt \qquad (7.5.3).$$

<u>Proof.</u> We will show that for each $1 \le k \le n-1$

$$\pi n^{-1} \int_{k\pi n^{-1}}^{(k+1)\pi n^{-1}} t^{-2}\,\omega(t)\,dt \le k^{-2}\,\omega(n^{-1}(k+1)\pi) \qquad (7.5.4)$$

$$\le 8\pi n^{-1} \int_{k\pi n^{-1}}^{(k+1)\pi n^{-1}} t^{-2}\omega(t)dt$$

which clearly implies (7.5.3).

Using property (1.3.4), of the modulus of continuity, we find that when $n^{-1}k\pi \le t \le n^{-1}(k+1)\pi$, then

$$n^2\pi^{-2}(k+1)^{-2}\omega(n^{-1}(k+1)\pi) \le 2t^{-2}\omega(t) \quad .$$

So after integrating, and using the fact that $k^{-2} \le 4(k+1)^{-2}$, we find

$$(4\pi)^{-1}nk^{-2}\omega(n^{-1}(k+1)\pi) \le n\pi^{-1}(k+1)^{-2}\omega(n^{-1}(k+1)\pi) \le 2 \int_{k\pi n^{-1}}^{(k+1)\pi n^{-1}} t^{-2}\omega(t)\,dt$$

which gives the right half of (7.5.4).

For the other estimate, we need only use the monotonicity of ω to conclude that

$$\int_{k\pi n^{-1}}^{(k+1)\pi n^{-1}} t^{-2}\,\omega(t)dt \le \omega(n^{-1}(k+1)\pi) \int_{k\pi n^{-1}}^{(k+1)\pi n^{-1}} t^{-2}\,dt$$

$$\le n\pi^{-1}\,k^{-2}\,\omega(n^{-1}(k+1)\pi)$$

as desired.

<u>Proof of Theorem 7.10.</u> First of all, if $f \in C_\omega(M)$ then for $x \in [-\pi,\pi]$

$$|\sigma_n(f,x) - f(x)| = \left|\frac{1}{2\pi}\int_{-\pi}^{\pi} (f(x+t) + f(x-t) - 2f(x))\,F_n(t)\,dt\right|$$

$$\le \frac{M}{\pi}\int_{-\pi}^{\pi} \omega(|t|)\,F_n(t)dt$$

$$= \frac{2M}{\pi}\int_{0}^{\pi} \omega(t)\,F_n(t)\,dt \quad .$$

Also, the 2π-periodic extension of $\omega(|t|)$ is in $C_\omega(M)$. Therefore,

$$M^{-1}(E(C_\omega(M),\sigma_n)) = \frac{2}{\pi}\int_0^{\pi} \omega(t)\,F_n(t)$$

$$= \frac{2}{\pi}\int_0^{\pi(n+1)^{-1}} \omega(t)\,F_n(t)dt + \frac{2}{\pi}\int_{\pi(n+1)^{-1}}^{\pi} \omega(t)\,F_n(t)dt$$

$$= I_1(n) + I_2(n) \qquad (7.5.5).$$

We first want to estimate $I_2(n)$. As usual, C will always denote a constant independent of n. It will be slightly more convenient to work with the index n-1 instead of n . To estimate from above, we need only observe that if $0 \leq t \leq \pi$ then

$$F_{n-1}(t) = \frac{1}{2n}\left(\frac{\sin(\frac{nt}{2})}{\sin\frac{t}{2}}\right)^2 \leq \frac{1}{2}\pi^2 n^{-1} t^{-2}$$

so that

$$I_2(n-1) \leq Cn^{-1} \int_{\pi n^{-1}}^{\pi} t^{-2}\,\omega(t)dt \qquad (7.5.6).$$

We can get a similar estimate from below. Let

$$J_k = [\pi k n^{-1} + \pi(4n)^{-1}\ ,\ \pi k n^{-1} + 3\pi(4n)^{-1}] \quad .$$

For $1 \leq k \leq n-1$ and $t \in J_k$, we have

$$\left(\frac{\sin(\frac{nt}{2})}{\sin\frac{t}{2}}\right)^2 \geq (4\sin^2\frac{\pi}{8})t^{-2} \quad .$$

Therefore, using the monotonicity of ω , and our last inequality, we find

$$2\pi^{-1}\int_{k\pi n^{-1}}^{(k+1)\pi n^{-1}} \omega(t)F_{n-1}(t)dt \geq \frac{1}{n\pi}\int_{J_k} \omega(t) \left(\frac{\sin(\frac{nt}{2})}{\sin\frac{t}{2}}\right)^2 dt$$

$$\geq Cn^{-1}\int_{J_k} t^{-2}\omega(t)\,dt \quad \geq Ck^{-2}\omega(n^{-1}k\pi)$$

$$\geq Ck^{-2}\,\omega(n^{-1}(k+1)\pi)$$

with C a constant independent of n . Now, we sum over k and envoke Lemma 7.1 to obtain

$$I_2(n-1) \geq C\,n^{-1}\int_{\pi n^{-1}}^{\pi} t^{-2}\,\omega(t)\,dt \qquad (7.5.7).$$

We change variables in the last integral to find

$$\int_{\pi n^{-1}}^{\pi} t^{-2}\,\omega(t)dt = \frac{1}{\pi}\int_1^n \omega(\pi t^{-1})dt \quad .$$

Hence, using the monotonicity of ω as well as property (1.3.3) of ω , we find that if $t \in [k,k+1]$ then $\omega(k^{-1}) \leq \omega(\pi(k+1)^{-1}) \leq \omega(\pi t^{-1}) \leq (1+\pi)\omega(k^{-1})$. Thus

$$\sum_{k=1}^{n-1}\omega(k^{-1}) \leq \int_{\pi n^{-1}}^{\pi} t^{-2}\,\omega(t)dt \leq (1+\pi)\sum_{k=1}^{n-1}\omega(k^{-1}) \quad .$$

This coupled with our estimates (7.5.6) and (7.5.7) shows that

$$C_3 n^{-1} \sum_{k=1}^{n-1} \omega(k^{-1}) \le I_2(n-1) \le C_4 n^{-1} \sum_{k=1}^{n-1} \omega(k^{-1}) \qquad (7.5.8)$$

with C_3 and C_4 constants independent of n.

Finally, since $\omega(k^{-1}) \ge \omega(n^{-1})$ for $k \le n$, we have

$$\pi I_1(n-1) = 2 \int_0^{\pi n^{-1}} \omega(t) F_n(t)dt \le \omega(\pi n^{-1}) \le (1+\pi)\omega(n^{-1})$$

$$\le 2(1+\pi)n^{-1} \sum_{k=1}^{n-1} \omega(k^{-1}) \quad .$$

This together with (7.5.8) shows that for certain constants $C_1, C_2 > 0$, we have

$$C_1 n^{-1} \sum_{k=1}^{n-1} \omega(k^{-1}) \le I_1(n-1) + I_2(n-1) \le C_2 n^{-1} \sum_{k=1}^{n-1} \omega(k^{-1}) \quad .$$

This gives the desired inequality because of (7.5.5).

As a second example, let (H_n) be the sequence of Hermite-Fejér operators which were introduced in Section 2.1. The following theorem of R. Bojanic[2] gives the weak asymptotic behavior of $E(C_\omega(M), H_n)$. This result also gives a considerable improvement over the estimates for the degree of approximation by (H_n) given in Section 2.8.

Theorem 7.11. _There exists constants_ $C_1, C_2 > 0$ _such that for any modulus of continuity_ ω _on_ $[0,2]$

$$C_1 M n^{-1} \sum_{k=1}^{n} \omega(k^{-1}) \le E(C_\omega(M), H_n) \le C_2 M n^{-1} \sum_{k=1}^{n} \omega(k^{-1}) \qquad (7.5.9).$$

Proof: Arguing as we have in Theorem 7.7 for the special case $\omega(t) = t^{\alpha}$ shows that in general

$$M^{-1}(E(C_{\omega}(M),H_n)) = ||H_n(\omega(|t-x|),x)||$$

$$= ||\sum_{k=1}^{n} \omega(|x_{k,n}-x|)(1-xx_{k,n}) \left(\frac{T_n(x)}{n(x-x_{k,n})}\right)^2|| \qquad (7.5.10)$$

where the norm is taken over $x \in [-1,1]$. No "o" term appears because H_n preserves constants.

Most of the estimates that follow are similar to or exactly the same as the estimates used in the previous proof, so at various times we will take the liberty of refering the reader back to Theorem 7.10 for details.

We first estimate from above by transforming to the trigonometric case. Suppose $x \in [-1,1]$ and write $x = \cos\theta$, $0 \le \theta \le \pi$. Now,

$$x_{k,n} = \cos\theta_{k,n}, \theta_{k,n} = (k - \tfrac{1}{2})\pi n^{-1} \quad , \quad k = 1,\cdots,n \quad .$$

The function

$$\frac{1 - \cos x \cos y}{(\cos x - \cos y)^2} \sin^2(\frac{x-y}{2})$$

is continuous on the rectangle $[o,\pi]\times[o,\pi]$ and hence bounded in this rectangle.

Since

$$\cos^2 n\theta = \sin^2 n(\theta-\theta_{k,n}) \le 4 \sin^2 n\left(\frac{\theta-\theta_{k,n}}{2}\right) \quad ,$$

we can estimate

$$n^{-2}(1 - \cos\theta \cos\theta_{k,n})\left(\frac{\cos n\theta}{\cos\theta - \cos\theta_{k,n}}\right)^2 \leq C_3 n^{-1} F_{n-1}(\theta-\theta_{k,n}) \quad (7.5.11)$$

with F_{n-1} the Fejér Kernel, and C_3 a constant. Now, fix θ and note that $|\cos\theta - \cos\theta_{k,n}| \leq |\theta-\theta_{k,n}|$, so that (7.5.10) leads us to estimate

$$n^{-1} \sum_{k=1}^{n} \omega(|\theta-\theta_{k,n}|)\, F_{n-1}(\theta-\theta_{k,n}) = \sum{}' + \sum{}'' \quad (7.5.12)$$

where $\sum'$ indicates the sum over those indices k for which $|\theta-\theta_{k,n}| \leq \pi n^{-1}$ and $\sum''$ the remaining sum. In the expression (7.5.12), the variable $|\theta-\theta_{k,n}|$ can be larger than 2 so that we consider the modulus of continuity to be defined for $x > 2$ by $\omega(x) = \omega(2)$.

Now, $F_{n-1}(t) \leq n$ for all t and there are at most two indices involved in the sum $\sum'$, so that

$$\sum{}' \leq 2\omega(\pi n^{-1}) \leq 2(1+\pi)\, \omega(n^{-1}) \quad (7.5.13).$$

For the second sum, if $t \in [k\pi n^{-1}, (k+1)\pi n^{-1}]$, then

$$F_{n-1}(t) \leq \frac{1}{2n}\left(\frac{\sin(\frac{nt}{2})}{\sin\frac{t}{2}}\right)^2 \leq \frac{1}{2}\pi^2 n^{-1} t^{-2} \leq \frac{1}{2} n k^{-2} .$$

Again, for each $k \geq 1$ there are at most two indices j with

$$k^{-1} n\pi < |\theta-\theta_{j,n}| \leq (k+1) n^{-1}\pi$$

and so

$$\sum{}'' \leq \sum_{k=1}^{n-1} k^{-2}\, \omega(n^{-1}(k+1)\pi) \quad (7.5.14).$$

Our estimates on $\sum'$ and $\sum''$ allow us to conclude that

$$M^{-1}(E(C_\omega(M),H_n)) = \left\| \sum_{k=1}^{n} \omega(|x_{k,n}-x|)(1 - xx_{k,n})\left(\frac{T_n(x)}{n(x-x_{k,n})}\right)^2 \right\| [-1,1]$$

$$\leq C_3 n^{-1} \left\| \sum_{k=1}^{n} \omega(|\theta_{k,n}-\theta|)\, F_{n-1}(\theta-\theta_{k,n}) \right\| [0,\pi]$$

$$\leq 2C_3(1+\pi)\ \omega(n^{-1}) + C_3 \sum_{k=1}^{n-1} k^{-2}\ \omega(n^{-1}(k+1)\pi) \qquad (7.5.15)$$

where we have used (7.5.11) to obtain the first inequality. Now, we use Lemma 7.1 on (7.5.15) to find

$$M^{-1}(E(C_\omega(M),H_n)) \leq C_4(\omega(n^{-1}) + n^{-1} \int_{\pi n^{-1}}^{\pi} t^{-2}\ \omega(t)dt) \quad .$$

Making a change of variables, and arguing exactly as we did in the proof of Theorem 7.9, we find

$$M^{-1}(E(C_\omega(M),H_n) \leq C_2 n^{-1} \sum_{k=1}^{n-1} \omega(k^{-1}) \leq C_2 n^{-1} \sum_{k=1}^{n} \omega(k^{-1}) \qquad (7.5.16)$$

which is the right hand side of (7.5.9).

To establish the lower estimate in (7.5.9), we take a particular value for x . When n is even we can take $x = 0$, when n is odd we can take $x = \cos(\frac{n-1}{2n}\pi)$. For example, when n is even then $(T_n(0))^2 = 1$ and so using (7.5.10), we find

$$M^{-1}(E(C_\omega(M),H_n)) \geq n^{-2} \sum_{k=1}^{n} \omega(|x_{k,n}|)(\frac{T_n(0)}{x_{k,n}})^2 = n^{-2} \sum_{k=1}^{n} x_{k,n}^{-2}\ \omega(|x_{k,n}|) \quad (7.5.17).$$

We write $k = \frac{1}{2}n + j$, $-\frac{1}{2}n \leq j < \frac{1}{2}n$. Then, $|x_{k,n}| \leq (|j|+1)\pi n^{-1}$ and so because of property (1.3.4) of the modulus of continuity

$$2|x_{k,n}|^{-1}\ \omega(|x_{k,n}|) \geq n\pi^{-1}(|j|+1)^{-1}\ \omega(\pi(|j|+1)n^{-1}) \quad .$$

This means that

$$2x_{k,n}^{-2}\ \omega(|x_{k,n}|) \geq n^2\ \pi^{-2}(|j|+1)^{-2}\ \omega(\pi(|j|+1)n^{-1}) \quad .$$

Using this in (7.5.17) gives

$$M^{-1}(E(C_\omega(M),H_n)) \geq \frac{1}{2}\ \pi^{-2} \sum_{j=-\frac{1}{2}n}^{\frac{1}{2}n} (|j|+1)^{-2}\ \omega(\pi(|j|+1)n^{-1})$$

$$\geq C_5 \sum_{k=1}^{n} k^{-2}\ \omega(\pi(k+1)n^{-1}) \qquad (7.5.18).$$

In the last inequality, we have used property (1.3.3) and the following estimates, which holds for $1 \leq k \leq \frac{n}{2}$

$$(2k)^{-2}\ \omega(\pi(2k+1)n^{-1}) \leq \frac{k^{-2}}{4}\ \omega(3\pi kn^{-1}) \leq k^{-2}\ \omega(\pi(k+1)n^{-1})$$

and

$$(2k+1)^{-2}\ \omega(\pi(2k+2)n^{-1}) \leq \frac{k^{-2}}{4}\ \omega(4\pi kn^{-1}) \leq k^{-2}\ \omega(\pi kn^{-1}) \quad .$$

To complete the proof we use Lemma 7.1 in (7.5.18) to obtain

$$M^{-1}(E(C_\omega(M),H_n)) \geq C_6 n^{-1} \int_{\pi n^{-1}}^{\pi} t^{-2}\ \omega(t)dt \quad .$$

and then argue exactly as we did in Theorem 7.9 to conclude that

$$M^{-1}(E(C_\omega(M),H_n) \geq C_1 n^{-1} \sum_{k=1}^{n} \omega(k^{-1})$$

as desired.

7.6. Notes.

7.6.1 We have only developed the fundamental results in the theory of approximation of classes of functions. A good source for the state of this area is the article of E.Gorlich and E. $Stark_1$. This article includes a historical survey of most of the subject with references to 138 papers. For some more recent results, the reader should consult the paper of R.J. $Nessel_1$.

7.6.2 As we have remarked, what we have called a strong asymptotic analysis is not the strongest analysis possible. Asymptotic expansions with several terms are more difficult to obtain but have been given in certain cases. For example, P.L. Butzer and E. $Stark_1$ have given a five term asymptotic expansion for $E(Lip^*1,\sigma_n)$, the error in approximating Lip^*1 by the Fejér operators. Subsequently S. $Teljakovskii_1$ found the complete asymptotic expansion for $E(Lip^*1,\sigma_n)$. The analogous result for approximation by the Abel-Poisson operators was given by E. $Stark_4$.

7.6.3 In Theorem 7.7, we were able to give a strong asymptotic analysis of $E(Lip(\alpha,M),L_n)$, $0 < \alpha \leq 1$, but only a weak analysis for $E(W^{(1)}(\alpha-1,\alpha,M),L_n)$, $1 < \alpha \leq 2$. The reason we could only get a weak analysis in the latter case is that the function $g(t) = |t|^\alpha$ is not in the class $W^{(1)}(\alpha-1,\alpha,M)$. The derivative of g is $g'(t) = t^{\alpha-1}$ sgn t and hence $g'(\frac{t}{2}) - g'(-\frac{t}{2}) = 2^{2-\alpha}|t|^{\alpha-1} > |t|^{\alpha-1}$. The function $2^{\alpha-2}g(t)$ is in $W^{(1)}(\alpha-1,\alpha,M)$, which enabled us to get the weak analysis.

CHAPTER 8

Inverse Theorems

8.1. Introduction. The most precise way of measuring the sharpness of direct estimates is by inverse theorems. An inverse theorem is a result which infers smoothness from the degree of approximation.

Suppose A is a class of functions and (L_n) is a sequence of linear operators for which the weak asymptotic behavior $(E(A,L_n)) \sim (\phi_n)$ is known. This subsumes a direct estimate, namely, $f \in A$ implies $||f-L_n(f)|| = O(\phi_n)$. This direct estimate cannot be improved for the class A in the sense that (ϕ_n) cannot be replaced by a sequence which tends to 0 faster than (ϕ_n).

A more exact relation between the smoothness and the degree of approximation occurs when we can show that $||f-L_n(f)|| = O(\phi_n)$ implies $f \in A$. This is an inverse theorem for the class A. It is clear that in such an analysis, we want the direct and inverse theorems for A to be given in terms of the same sequence (ϕ_n). Therefore, when $(E(A,L_n))$ is known asymptotically, we must take $(\phi_n) \sim (E(A,L_n))$.

Inverse theorems are generally much more difficult to obtain than estimates for the approximation of classes of functions. In fact, for a given class of functions A, it is not necessarily true that $||f-L_n(f)|| = O(E(A,L_n))$ implies $f \in A$. Certainly, this is always the case for a proper subclass of the saturation class.

As another example, for (σ_n) the sequence of Fejér operators and $A = \text{Lip } 1$, then $(E(A,\sigma_n)) \sim (n^{-1}\log n)$. However, there are functions in $S(\sigma_n)$ which are approximated with order $(n^{-1} \log n)$ (in fact order (n^{-1})) but are not in $\text{Lip } 1$.

We can preview at least one of the difficulties that arises in trying to prove inverse theorems by considering the following simple sequence of operators. Let $d\mu_n = \frac{\pi}{2}(d\rho_{t_n} + d\rho_{-t_n})$ where $t_n \downarrow 0$. Consider the sequence of operators (L_n) defined on C^* by $L_n(f) = f * d\mu_n$. The sequence (L_n) is saturated with order (t_n^2) and saturation class Lip^*2 (see Theorem 3.6). Also, for each $0 < \alpha \leq 2$, $E(\mathrm{Lip}^*(\alpha,M),L_n) \sim (t_n^\alpha)$. This is a consequence of Theorem 7.4.

In the saturation case, $\alpha = 2$, it follows that if

$$\frac{1}{2}||f(x+t_n) + f(x-t_n) - 2f(x)|| = ||f-L_n(f)|| = 0(t_n^2) \qquad (8.1.1)$$

then $f \in \mathrm{Lip}^*2$. It is generally not possible to replace t_n^2 by t_n^α in (8.1.1) and still conclude that $f \in \mathrm{Lip}^*\alpha$. In fact, we will show that if $0 < \alpha < 2$, then in order to have

$$||f(x+t_n) + f(x-t_n) - 2f(x)|| = 0(t_n^\alpha) \qquad (8.1.2)$$

always imply $f \in \mathrm{Lip}^*\alpha$, it is necessary and sufficient that there is a constant $C > 0$ for which

$$\frac{t_n}{t_{n+1}} \leq C \qquad (8.1.3).$$

This indicates that in order to prove general inverse theorems, we will need some restriction on how fast the saturation order tends to 0, or viewed in another way, we will need to assure that the sequence of operators is not too sparse.

Let us demonstrate the necessity of (8.1.3). The sufficiency will follow from some general considerations in Section 8.3. The following argument was kindly pointed out to me by Professor H.S. Shapiro.

Let $0 < \alpha < 2$ and define the following two semi-norms

$$||f||^*_{\alpha,1} = \sup_{t>0} t^{-\alpha} ||f(x+t) + f(x-t) - 2f(x)||$$

$$||f||^*_{\alpha,2} = \sup_{n} t_n^{-\alpha} ||f(x+t_n) + f(x-t_n) - 2f(x)|| \qquad (8.1.4).$$

We suppose that (8.1.3) does not hold and wish to show that there is a function f with $||f||^*_{\alpha,1} = \infty$ and $||f||^*_{\alpha,2} < \infty$. We first show that there is a sequence $\epsilon_n \downarrow 0$ such that for each n we can find functions f_n with

$$\epsilon_n^{-1} < ||f_n||^*_{\alpha,2} < \epsilon_n ||f_n||^*_{\alpha,1} \qquad (8.1.5).$$

Let $\delta_n = t_{n+1} t_n^{-1}$, so that δ_n can be chosen arbitrarily small by a suitable choice of n. Define $f_n(x) = e^{2\pi i\lambda_n x}$ where $\lambda_n = [\delta_n^{-1/2} t_n^{-1}]$. First of all,

$$||f_n||^*_{\alpha,1} \geq (2\lambda_n)^{\alpha} |f_n((2\lambda_n)^{-1}) + f_n(-(2\lambda_n)^{-1}) - 2f_n(0)| = 2^{2+\alpha}\lambda_n^{\alpha} \qquad (8.1.6).$$

To estimate $||f_n||^*_{\alpha,2}$ from above, we first consider the case when $k \leq n$, so that

$$t_k^{-\alpha} ||f_n(x+t_k) + f_n(x-t_k) - 2f_n(x)|| \leq 4t_k^{-\alpha} \leq 4t_n^{-\alpha} \leq 4(\lambda_n+1)^{\alpha} \delta_n^{\alpha/2} \qquad (8.1.7).$$

When $k > n$, we have

$$t_k^{-\alpha}\,||f_n(x+t_k)+f_n(x-t_k)-2f_n(x)|| = t_k^{-\alpha}\,|e^{2\pi i\lambda_n t_k}+e^{-2\pi i\lambda_n t_k}-2|$$

$$\leq 4t_k^{-\alpha}\sin^2(\pi\lambda_n t_k)$$

$$\leq 4\pi^2(t_k\lambda_n)^{2-\alpha}\lambda_n^{\alpha}$$

$$\leq 4\pi^2(t_{n+1}\lambda_n)^{2-\alpha}\lambda_n^{\alpha}$$

$$\leq 4\pi^2\,\delta_n^{1-\alpha/2}\,\lambda_n^{\alpha} \qquad (8.1.8).$$

The estimates (8.1.7) and (8.1.8) show that

$$||f_n||_{\alpha,2}^{*} \leq 4\pi^2(\lambda_n+1)^{\alpha}\,(\delta_n^{\alpha/2}+\delta_n^{1-\alpha/2}) \qquad (8.1.9).$$

Since δ_n can be chosen arbitrarily small, which in turn makes λ_n large, (8.1.6) and (8.1.9) can be used to show (8.1.5).

Now, we define $||f||_{\alpha,1} = ||f||_{\alpha,1}^{*} + ||f||$ and $||f||_{\alpha,2} = ||f||_{\alpha,2}^{*} + ||f||$. Then, $||\cdot||_{\alpha,1}$ and $||\cdot||_{\alpha,2}$ are both norms on $\mathrm{Lip}^{*}\alpha$. Also, $\mathrm{Lip}^{*}\alpha$ is complete under $||\cdot||_{\alpha,1}$. It is clear that if $f \in \mathrm{Lip}^{*}\alpha$, then

$$||f||_{\alpha,2} \leq ||f||_{\alpha,1} \qquad (8.1.10).$$

Because of (8.1.5), $||\cdot||_{\alpha,1}$ can not be dominated by a multiple of $||\cdot||_{\alpha,2}$ (note that $||f_n|| \leq 1$, for each n). This means that $\mathrm{Lip}^{*}\alpha$ cannot be complete under $||\cdot||_{\alpha,2}$, (See Dunford and Schwartz$_1$, pg. 58). Hence, there is a sequence of functions (g_n) from $\mathrm{Lip}^{*}\alpha$ which is Cauchy in $||\cdot||_{\alpha,2}$ but has no limit in $\mathrm{Lip}^{*}\alpha$. The sequence (g_n) is also Cauchy in $||\cdot||$, so there is a function $g \in C^*$ such that $||g-g_n|| \to 0$. We can check that (g_n) converges to g in $||\cdot||_{\alpha,2}$. Since, if $\varepsilon > 0$, choose N so that n_1, $n_2 \geq N$ imply $||g_{n_1} - g_{n_2}||_{\alpha,2} < \varepsilon$. Then for each m,

$$t_m^{-\alpha} ||g_{n_1}(x+t_m) + g_{n_1}(x-t_m) - 2g_{n_1}(x) - [g(x+t_m) + g(x-t_m) - 2g(x)]||$$

$$= \lim_{n_2\to\infty} t_m^{-\alpha} ||g_{n_1}(x+t_m) + g_{n_1}(x-t_m) - 2g_{n_1}(x) - [g_{n_2}(x+t_m) + g_{n_2}(x-t_m) - 2g_{n_2}(x)]||$$

$$< \epsilon .$$

This shows that for $n_1 \geq N$

$$\sup_m t_m^{-\alpha} ||g_{n_1}(x+t_m) + g_{n_1}(x-t_m) - 2g_{n_1}(x) - (g(x+t_m) + g(x-t_m) - 2g(x))|| < \varepsilon$$

as desired. By what we have remarked, g cannot be in $\mathrm{Lip}^*\alpha$, so that $||g||_{\alpha,1} = \infty$, while

$$||g||_{\alpha,2} \leq \sup_n ||g_n||_{\alpha,2} < +\infty .$$

This shows that (8.1.3) is a necessary condition.

8.2. Inverse Theorems for Operators Which Satisfy Bernstein Type Inequalities

We already know inverse theorems for certain sequences of polynomial operators (see Chapters 4 and 6 for the definition of a sequence of polynomial operators). These are obtained from the classical inverse theorems of Bernstein, Zygmund, etc., for approximation by polynomials (Section 1.5). For example, if (L_n) is an optimal sequence of trigonometric operators, then for $0 < \alpha < 2$, $f \in \mathrm{Lip}^*\alpha$ if and only if

$$||f - L_n(f)|| = O(n^{-\alpha}) .$$

The proofs of these inverse theorems rest on the Bernstein and Markov inequalities for derivatives of polynomials (see G.G. Lorentz$_2$, pg. 39). The techniques involved in the proof are very important because of their wide application to general inverse theorems. For example, in P.L. Butzer and K. Scherer$_1$ they have developed a general theory for obtaining inverse theorems for certain sequences of operators defined on a Banach space. In this section, we want to illustrate at least the fundamental ideas involved by considering the special spaces $Lip^*\alpha$.

We suppose that (L_n) is a sequence of operators (not necessarily positive) defined on $C^*[-\pi,\pi]$ (an analogous development can be given for $C[a,b]$). We also require that L_n commutes with L_m for each n and m . This latter condition can be avoided in certain cases, for example when (L_n) is a sequence of polynomial operators. The prototype example which satisfies the requirements is a sequence of convolution operators.

Let $0 < \alpha \leq 2$. Then $Lip^*\alpha$, becomes a Banach space under the norm $||\cdot||_{\alpha,1}$ introduced in Section 8.1. Suppose for some $C_1 > 0$, we have the direct estimate

$$||f - L_n(f)|| \leq C_1 \, ||f||_{\alpha,1} \, \phi_n^{\alpha} \tag{8.2.1}$$

for each $f \in Lip^*\alpha$. Whenever $E(Lip^*(\alpha,M),L_n)$ is known asymptotically, such an estimate is available.

We say that (8.2.1) has a companion Bernstein type inequality, if there is a constant $C_2 > 0$ such that for each $f \in C^*$

$$||L_n(f)||_{\alpha,1} \leq C_2 \, \phi_n^{-\alpha} \, ||f|| \tag{8.2.2}.$$

This inequality takes the role of the classical Bernstein inequality for derivatives of polynomials in the Bernstein inverse theorem. Under these conditions, we can give an inverse theorem for the class $\mathrm{Lip}^*\beta$ whenever $\beta < \alpha$. First, we must give the following direct theorem since the direct estimates we need do not follow from our previous results. This theorem shows how estimates for $\mathrm{Lip}^*\alpha$ always provide estimates for $\mathrm{Lip}^*\beta, \beta < \alpha$, which in some instances (e.g. Fejér operators) are better than the estimates given in Chapter 2 (see also 2.9.3 and Chapter 7).

Theorem 8.1. Let (L_n) be a bounded sequence of linear operators defined on C^*. If for some $0 < \alpha \leq 2$, (L_n) satisfies (8.2.1), then for each $\beta \leq \alpha$ we have

$$||f - L_n(f)|| = O(\phi_n^\beta)$$

whenever $f \in \mathrm{Lip}^*\beta$.

Remark. We do not assume that the operators are positive or that they commute.

Proof. Suppose $f \in \mathrm{Lip}^*(\beta,M)$ (see Section 1.3 for notation). For $\varepsilon > 0$, let

$$f_\varepsilon(x) = \varepsilon^{-2} \int_{-\varepsilon/2}^{\varepsilon/2} \int_{-\varepsilon/2}^{\varepsilon/2} f(x+u+v)\,du\,dv .$$

Estimating as we have in the proof of Lemma 6.7, we can show that

$$||f - f_\varepsilon|| \leq \omega_2(f,\varepsilon) \qquad (8.2.3)$$

$$||f_\varepsilon''|| \leq \varepsilon^{-2}\omega_2(f,\varepsilon) \qquad (8.2.4)$$

$$\omega_2(f_\varepsilon,t) \leq \omega_2(f,t) \qquad 0 < t \leq \pi \qquad (8.2.5).$$

We take $\varepsilon = \phi_n$. For $t \leq \phi_n$, it follows from (8.2.4) that

$$t^{-\alpha} ||f_{\phi_n}(x+t) + f_{\phi_n}(x-t) - 2f_{\phi_n}(x)|| \leq t^{2-\alpha} ||f''_{\phi_n}|| \leq \phi_n^{-2} \omega_2(f,\phi_n)t^{2-\alpha}$$

$$\leq 2M\phi_n^{-2+\beta} \phi_n^{2-\alpha} \leq 2M\phi_n^{\beta-\alpha} .$$

When $t \geq \phi_n$, we use (8.2.5) to see that

$$t^{-\alpha} ||f_{\phi_n}(x+t) + f_{\phi_n}(x-t) - 2f_{\phi_n}(x)|| \leq t^{-\alpha} 2Mt^{\beta} \leq 2M\phi_n^{\beta-\alpha} .$$

Thus,

$$||f_{\phi_n}||_{\alpha,1} \leq 2M\phi_n^{\beta-\alpha} + ||f_{\phi_n}|| \leq 2M\phi_n^{\beta-\alpha} + ||f|| \qquad (8.2.6).$$

Now, we can estimate

$$||f - L_n(f)|| \leq ||f-f_{\phi_n}|| + ||f_{\phi_n} - L_n(f_{\phi_n})|| + ||L_n(f-f_{\phi_n})||$$

$$\leq (1 + ||L_n||) ||f-f_{\phi_n}|| + ||f_{\phi_n} - L_n(f_{\phi_n})||$$

$$\leq (1 + ||L_n||) 2M\phi_n^{\beta} + C_1||f_{\phi_n}||_{\alpha,1} \phi_n^{\alpha} = O(\phi_n^{\beta})$$

where $||f-f_{\phi_n}||$ was estimated by (8.2.3) and $||f_{\phi_n} - L_n(f_{\phi_n})||$ was estimated by using (8.2.1) and (8.2.6).

Theorem 8.2. Let (L_n) be a bounded sequence of linear operators (not necessarily positive) from C^* to C^* such that for each n and m, L_n commutes with L_m. Suppose that for some $0 < \alpha \leq 2$, (8.2.1) and (8.2.2) hold for a sequence

(ϕ_n) <u>which converges to</u> 0 <u>and satisfies</u>

$$\frac{\phi_n}{\phi_{n+1}} \leq \delta < +\infty \quad , \quad n \in N \tag{8.2.7}$$

<u>Then, for</u> $0 < \beta < \alpha$, <u>a function</u> f <u>is in</u> $\text{Lip}^*\beta$ <u>if and only if</u>

$$||f - L_n(f)|| = O(\phi_n^\beta) \tag{8.2.8}$$

<u>Remarks</u>. The theorem gives no information for $\beta = \alpha$. This is usually the saturation case and is solved by other methods.

<u>Proof</u>. The direct part of the theorem is a consequence of Theorem 8.1. For the inverse direction, suppose f satisfies (8.2.8). We want to show that $||\Delta_t^2 (f)|| = O(t^\beta)$. Choose a subsequence (n_k) such that

$$2\phi_{n_{k+1}} \leq \phi_{n_k} \tag{8.2.9}$$

$$\phi_{n_k} \leq 2\delta\phi_{n_{k+1}} \tag{8.2.10}$$

Note that $\delta > 1$, since (ϕ_n) converges to 0 . Such a sequence is given by defining $n_1 = 1$ and inductively

$$n_{k+1} = \min\{n : n > n_k \ , \ 2\phi_n < \phi_{n_k}\} \ .$$

Condition (8.2.9) guarantees that $\phi_{n_{k+1}} \leq 2^{-k}\phi_{n_1}$.

Now, if $0 < t < \phi_{n_1}$ choose m so that $\phi_{n_{m+1}} \leq t < \phi_{n_m}$. Then, since

$$\sum_{k=1}^{\infty} \phi_{n_k}^\beta < +\infty \quad ,$$

we have

$$f = L_{n_1}(f) + \sum_{k=1}^{m} [L_{n_{k+1}}(f) - L_{n_k}(f)] + \sum_{k=m+1}^{\infty} [L_{n_{k+1}}(f) - L_{n_k}(f)]$$

and so

$$||\Delta_t^2(f)|| \leq ||\Delta_t^2 (L_{n_1}(f))|| + \sum_{k=1}^{m} ||\Delta_t^2 (L_{n_{k+1}}(f) - L_{n_k}(f))||$$
$$+ \sum_{k=m+1}^{\infty} ||\Delta_t^2 (L_{n_{k+1}}(f) - L_{n_k}(f))|| = I_1 + I_2 + I_3 \quad .$$

In the remaining estimates, C will always denote a constant depending only on α, β , and f . The estimate for I_1 is easy since

$$I_1 \leq t^\alpha ||L_{n_1}(f)||_{\alpha,1} \leq t^\alpha C_2 \phi_{n_1}^{-\alpha} ||f|| \leq Ct^\alpha \tag{8.2.11}$$

because of (8.2.2).

To estimate I_2 , we write

$$L_{n_{k+1}}(f) - L_{n_k}(f) = L_{n_{k+1}}(f - L_{n_k}(f)) - L_{n_k}(f - L_{n_{k+1}}(f))$$

because of commutivity. Therefore, from (8.2.2) and (8.2.8), we find

$$||\Delta_t^2(L_{n_{k+1}}(f) - L_{n_k}(f))|| \leq C_2\, t^\alpha\, \phi_{n_{k+1}}^{-\alpha}\, ||f - L_{n_k}(f)|| + C_2\, t^\alpha\, \phi_{n_k}^{-\alpha}\, ||f-L_{n_{k+1}}(f)||$$
$$\leq Ct^\alpha(\phi_{n_{k+1}}^{-\alpha}\, \phi_{n_k}^{\beta} + \phi_{n_k}^{-\alpha}\, \phi_{n_{k+1}}^{\beta})$$
$$\leq Ct^\alpha\, \phi_{n_k}^{-\alpha+\beta}$$

where in the last inequality we have used (8.2.10). Summing over k, we have

$$I_2 \le Ct^{\alpha} \sum_{k=1}^{m} \phi_{n_k}^{-\alpha+\beta} \le Ct^{\alpha} \phi_{n_m}^{\beta-\alpha} (1 + 2^{\beta-\alpha} + \cdots + 2^{m(\beta-\alpha)}) \le Ct^{\beta} \qquad (8.2.12)$$

where we have used (8.2.9) and the fact that $t \le \phi_{n_m}$.

Now to the estimate of I_3. For each $k > m$,

$$||\Delta_t^2(L_{n_{k+1}}(f) - L_{n_k}(f))|| \le 4(||f - L_{n_k}(f)|| + ||f - L_{n_{k+1}}(f)||) \le C\phi_{n_k}^{\beta} .$$

Therefore, because of (8.2.9), we have

$$I_3 \le C \sum_{k=m+1} \phi_{n_k}^{\beta} \le C\phi_{n_{m+1}}^{\beta} \sum_{k=1}^{\infty} 2^{-k\beta} \le C\phi_{n_{m+1}}^{\beta} \le Ct^{\beta} \qquad (8.2.13).$$

The estimates (8.2.11), (8.2.12), and (8.2.13) serve to show that

$$||\Delta_t^2 f|| = O(t^{\beta})$$

and so $f \in Lip^* \beta$, as desired.

For an application of Theorem 8.2, we let

$$v_n(t) = C_n \cos^{2n} \frac{t}{2}$$

with C_n the normalizing constant. The operators $L_n(f) = f*v_n$ were introduced by C. de la Valee Poussin. To evaluate C_n we make the change of variables $x = \cos^2 \frac{t}{2}$, and find

$$C_n^{-1} = \frac{1}{\pi}\int_{-\pi}^{\pi}\cos^{2n}\frac{t}{2}\,dt = \frac{2}{\pi}\int_0^{\pi}\cos^{2n}\frac{t}{2}\,dt = \frac{2}{\pi}\int_0^1 x^{n-1/2}(1-x)^{-1/2}\,dx$$

$$= \frac{2}{\pi}\,\frac{\Gamma(n+\frac{1}{2})\,\Gamma(\frac{1}{2})}{\Gamma(n+1)} = 2^{-2n+1}\,\frac{(2n)!}{(n!)^2} \qquad (8.2.14).$$

A similar estimate shows that

$$\frac{1}{\pi}\int_{-\pi}^{\pi}\sin^2\frac{t}{2}\,v_n(t)\,dt = \frac{2C_n}{\pi}\int_0^1 x^{n-1/2}(1-x)^{1/2}\,dx$$

$$= \frac{2C_n}{\pi}\,\frac{\Gamma(n+\frac{1}{2})\,\Gamma(\frac{3}{2})}{\Gamma(n+2)}$$

$$= (2n+2)^{-1} \qquad (8.2.15).$$

Theorem 7.4 of Chapter 7 gives the Jackson type inequality

$$||f - L_n(f)|| \le C_1\,||f||_{2,1}\,n^{-1} \qquad (8.2.16)$$

where $||\cdot||_{2,1}$ is the norm introduced for $\dot{L}ip^{*}2$ in Section 8.1.

For the Bernstein type inequality, we need only realize that

$$v_n''(t) = n(n-\tfrac{1}{2})\,C_n\cos^{2n-2}\frac{t}{2}\sin^2\frac{t}{2} - \frac{1}{2}n\,v_n(t) \quad .$$

Therefore, for $n \ge 2$

$$\frac{1}{\pi}\int_{-\pi}^{\pi}|v_n''(t)|\,dt \le \frac{n^2C_n}{\pi}\int_{-\pi}^{\pi}\sin^2\frac{t}{2}\cos^{2n-2}\frac{t}{2}\,dt + \frac{n}{2\pi}\int_{-\pi}^{\pi}v_n(t)\,dt$$

$$= \frac{n^2\,C_n}{(2n)\,C_{n-1}} + \frac{n}{2} \le 2n$$

In the equality we used (8.2.15) and for the last inequality we used (8.2.14).

Now, if $f \in C^*$, then for the Lip^*2 semi-norm $(||\cdot||^*_{2,1})$, we have

$$||L_n(f)||^*_{2,1} \le ||(f*v_n)''|| = ||f*v_n''|| \le ||f|| \left(\frac{1}{\pi}\int_{-\pi}^{\pi} |v_n''|dt\right) \le 2||f||n .$$

Therefore

$$||L_n(f)||_{2,1} \le 2||f||n + ||f|| \le 3n||f|| \tag{8.2.17}$$

which is the companion inequality to (8.2.16). Using Theorem 8.2 with $\alpha = 2$ gives the following corollary.

<u>Corollary 8.1</u>. <u>For the de la Valee Poussin operators</u> $L_n(f) = f*v_n$, <u>and</u> $0 < \alpha \le 2$, <u>we have</u> $f \in \mathrm{Lip}^*\alpha$ <u>if and only if</u>

$$||f - L_n(f)|| = O(n^{-\alpha/2}) \quad .$$

<u>Proof</u>. Theorem 8.2 shows that the direct result holds. For the inverse direction, for $0 < \alpha < 2$, this is again a consequence of Theorem 8.2, since we have verified the Bernstein type inequality in (8.2.17). For $\alpha = 2$, the saturation result, we can use Tureckiĭ's theorem (Theorem 3.6). A similiar estimate to (8.2.15) shows that

$$\frac{1}{\pi}\int_{-\pi}^{\pi} \sin^4 \frac{t}{2}\, v_n(t)dt = O(n^{-2}) = o(n^{-1})$$

which is (3.9.3). This is equivalent to the hypotheses in Tureckiĭ's theorem.

<u>8.3. Inverse Theorems for Positive Convolution Operators.</u> The remainder of this chapter will be devoted to developing a second technique for obtaining inverse theorems, which replaces the assumption that (L_n) satisfies a Bernstein type

inequality by certain restrictions on the concentration of "mass" of L_n. This technique applies only to positive operators. We will first discuss the case when (L_n) is a sequence of operators of convolution type from $C[-2a,2a]$, to $C[-a,a]$ by which we mean

$$L_n(f,x) = \frac{1}{2}\int_{-a}^{a} (f(x+t)+f(x-t))d\mu_n(t) \quad , \quad -a \leq x \leq a \qquad (8.3.1)$$

where $d\mu_n$ is a non-negative, even Borel, measure on $[-a,a]$ with $\int_{-a}^{a} d\mu_n(t) = 1$. In the periodic case, this reduces to our customary terminology of convolution operator, where we have incorporated the usual $\frac{1}{\pi}$ factor in the measure.

Let (ϕ_n) be defined by

$$\phi_n^2 = \int_{-a}^{a} t^2 \, d\mu_n(t) \qquad (8.3.2).$$

We assume throughout that (ϕ_n) converges to 0.

It is easy to see that if $f \in \mathrm{Lip}^*(\alpha,M)$ on $[-a,a]$, then $||f - L_n(f)|| = O(\phi_n^\alpha)$ since

$$||L_n(f) - f|| = ||\frac{1}{2}\int_{-a}^{a} (f(x+t) + f(x-t) - 2f(x))d\mu_n(t)||$$

$$\leq M\int_{-a}^{a} |t|^\alpha \, d\mu_n(t) \leq 2M\int_{0}^{\phi_n} |t|^\alpha \, d\mu_n(t)$$

$$+ 2M\phi_n^{\alpha-2}\int_{\phi_n}^{a} t^2 \, d\mu_n(t) \leq 2M\phi_n^\alpha \qquad (8.3.3).$$

As usual, the norm is in the image space and hence the supremum over $[-a,a]$. We want to examine under what conditions the converse result holds. Namely, when does $||f - L_n(f)|| = O(\phi_n^\alpha)$ imply $f \in \mathrm{Lip}^*\alpha$?

To establish such an inverse theorem, we will need a requirement about the concentration of "mass" of $d\mu_n$. Notice, that if for some $0 < \varepsilon < \alpha$, we had

$$\int_{-a}^{a} |t|^{\alpha-\varepsilon}\, d\mu_n(t) = O(\phi_n^{\alpha}) \tag{8.3.4}$$

then the inverse theorem could not hold. For, if $||f - L_n(f)|| = O(\phi_n^{\alpha})$, then we couldn't say $f \in \mathrm{Lip}^*\alpha$, since each function in $\mathrm{Lip}^*(\alpha-\varepsilon)$ already has this degree of approximation.

In order to prevent (8.3.4) from happening, we will impose the following condition on $(d\mu_n)$.

<u>8.3.5.</u> <u>For each</u> $\varepsilon > 0$, <u>there is an</u> A , <u>such that for</u> $n \in N$,

$$\int_{A\phi_n}^{a} t^2 d\mu_n(t) \le \varepsilon \int_{0}^{a} t^2 d\mu_n(t) = \frac{\varepsilon}{2}\phi_n^2 .$$

If (8.3.5) holds, then taking $\varepsilon = \frac{1}{2}$, we find for any $0 < \alpha \le 2$

$$\int_0^a |t|^{\alpha}\, d\mu_n(t) \ge \int_0^{A\phi_n} |t|^{\alpha}\, d\mu_n(t) \ge (A\phi_n)^{\alpha-2} \int_0^{A\phi_n} t^2\, d\mu_n(t)$$

$$\ge \frac{1}{4} A^{\alpha-2}\, \phi_n^{\alpha-2}\, \phi_n^2 = \frac{1}{4} A^{\alpha-2}\, \phi_n^{\alpha} \tag{8.3.6}$$

and thus (8.3.4) can not hold. Actually, (8.3.5) is not the weakest condition which will prevent (8.3.4) from happening, but it is not far off, as we will see from the following examples. For r a positive integer, let

$$d\mu_n = \frac{1}{2n} \sum_{k=1}^{n} (d\rho_{x_{k,n}} + d\rho_{-x_{k,n}})$$

on $[-\pi,\pi]$, where $x_{k,n} = n^{-r}k^{-1}$. Then

$$\phi_n^2 = \int_{-\pi}^{\pi} t^2 \, d\mu_n(t) = n^{-2r-1} \sum_1^n k^{-2} \sim n^{-2r-1}$$

and for $\beta > 1$

$$\int_{-\pi}^{\pi} |t|^{\beta} \, d\mu_n(t) = n^{-\beta r-1} \sum_1^n k^{-\beta} .$$

Thus, when $1 < \beta < 2$, there is an $\varepsilon > 0$ for which

$$\int_{-\pi}^{\pi} |t|^{\beta-\varepsilon} \, d\mu_n(t) \sim n^{-\beta r+\varepsilon r-1} = O(n^{-\beta r-\frac{1}{2}\beta})$$

so that the inverse theorem cannot hold.

However, for $(d\mu_n)$ we have

$$\int_{2n^{-r}}^{\pi} d\mu_n(t) = 0 \quad , \quad n \in N .$$

By taking r large, we see that it is not possible to weaken (8.3.5) much, since (n^{-r}) is close to (ϕ_n) .

It is important to point out that (8.3.5) is a much stronger condition than the conditions needed for the saturation theorem, which is the case when $\alpha = 2$ (see Theorems 3.10 and 5.4). In particular, in the periodic case, (8.3.5) guarantees that the asymptotic conditions of Theorem 3.8 hold. Therefore $\phi_n^2 \sim 1 - \rho_{1,n}$.

Our analysis will rest on refinements of the parabola technique introduced in Chapter 5.

<u>Lemma 8.1.</u> <u>Let</u> f <u>be continuous on</u> $[x_o-t_o, x_o+t_o]$, $t_o > 0$, <u>and suppose</u>

$$f(x_o+t_o) + f(x_o-t_o) - 2f(x_o) \leq -Mt_o^2 \qquad (8.3.7).$$

<u>If</u> $0 < \lambda < 1$, <u>then one of the following must hold:</u>

<u>8.3.8</u> <u>There is a point</u> $y \in [x_o-\lambda t_o, x_o+\lambda t_o]$ <u>and a quadratic</u> $Q^*(t) = -\frac{1}{2}(1-\lambda)M(t-y)^2 + \ell(t)$, ℓ <u>a linear function, with the properties</u>

$$Q^*(t) \geq f(t) \quad , \quad t \in [x_o-t_o, x_o+t_o]$$

$$Q^*(y) = f(y) \quad .$$

<u>8.3.9</u> <u>For some</u> $0 < \beta < 1-\lambda$, <u>and</u> $y \in [x_o-t_o, x_o+t_o]$

$$f(y+\beta t_o) + f(y-\beta t_o) - 2f(y) \leq -\frac{1}{2}(\lambda-2\beta-\beta^2)Mt_o^2 \quad .$$

<u>Proof.</u> Denote by ℓ_1 the linear function which interpolates f at x_o-t_o and x_o+t_o. The function $g = f-\ell_1$ vanishes at x_o-t_o and x_o+t_o and g also satisfies (8.3.7). Consider the quadratic $Q(t) = -\frac{1}{2}(1-\lambda)M(t-x_o)^2 + C$, where C is a constant chosen large enough to quarantee that $Q \geq g$ on $[x_o-t_o, x_o+t_o]$. Let

$$m = \inf_{t\in[x_o-t_o,x_o+t_o]} (Q(t) - g(t))$$

This infimum is attained at a point y in $[x_o-t_o, x_o+t_o]$. If $y \in [x_o-\lambda t_o, x_o+\lambda t_o]$, then (8.3.8) holds with $Q^* = 0 - m + \ell_1$.

Suppose then, that $y \notin [x_o-\lambda t_o, x_o+\lambda t_o]$. We want to show that (8.3.9) holds. We consider the case when $y = x_o-\lambda' t_o$, $\lambda' > \lambda$. The other case is handled similarily. It is easy to check that y is not either of the end points.

If we let $\beta = 1-\lambda'$ then $\beta > 0$ and

$$g(y-\beta t_o) = g(x_o-t_o) = 0 \tag{8.3.10}$$

In moving from the point x_o to the point y , the quadratic

$$Q(t) = -\frac{1}{2}(1-\lambda)\ M(t-x_o)^2 + C$$

decreases an amount

$$\frac{1}{2}(1-\lambda)M(\lambda' t_o)^2 \leq \frac{1}{2}(1-\lambda)\ Mt_o^2$$

and so

$$g(y) \geq \frac{1}{2}\lambda M t_o^2 \tag{8.3.11}$$

Otherwise, the infimum would be taken at x_o (note that $g(x_o) \geq \frac{1}{2} Mt_o^2$). Similarily, in moving from y to $y + \beta t_o$, Q changes an amount

$$|Q(y+\beta t_o) - Q(y)| = \frac{1}{2}(1-\lambda)M\ |2(y-x_o)\beta t_o + \beta^2\ t_o^2| \leq \frac{1}{2} M(2\beta+\beta^2)t_o^2$$

and so

$$g(y+\beta t_o) \leq g(y) + \frac{1}{2} M(2\beta+\beta^2)t_o^2 \tag{8.3.12}$$

When, we combine (8.3.10), (8.3.11), and (8.3.12), we see that

$$g(y-\beta t_o) + g(y+\beta t_o) - 2g(y) \leq -\frac{1}{2}(\lambda-2\beta-\beta^2)\ Mt_o^2 \quad .$$

Since $f = g + \ell_1$, f and g have the same second divided differences and thus (8.3.9) holds. Lemma 8.1 is proved.

Before proceeding to the statement and proof of our general inverse theorem, we return to the example of discretized second divided differences which were considered in Section 8.1. For $n \in N$, we let $d\mu_n = \frac{1}{2}(d\rho_{t_n} + d\rho_{-t_n})$, with $t_n \downarrow 0$. Because of the simple nature of the measures $d\mu_n$, the proof of the inverse theorem in this case is not complicated with technical difficulties. This will make the essential parts of the argument clearer.

We want to show that if $0 < \alpha \leq 2$ and $||f||^*_{\alpha,2}$ is finite then $||f||^*_{\alpha,1}$ is finite (see Section 8.1 for notation). We know from Section 8.1 that we must assume there is a constant $C > 0$ for which

$$\frac{t_n}{t_{n+1}} \leq C \quad , \qquad n \in N \tag{8.3.13}$$

which we do.

First, we will estimate $||f||^*_{\alpha,1}$ in terms of $||f||^*_{\alpha,2}$ whenever $||f||^*_{\alpha,1}$ is finite. So, suppose that $||f||^*_{\alpha,1} = M < +\infty$ and choose x_o , t_o such that

$$|f(x_o+t_o) + f(x_o-t_o) - 2f(x_o)| \geq \frac{9}{10} Mt_o^\alpha \ .$$

We can assume that

$$f(x_o+t_o) + f(x_o-t_o) - 2f(x_o) \leq -\frac{9}{10} M\, t_o^\alpha = -\frac{9}{10} M\, t_o^{\alpha-2}\, t_o^2$$

by considering $-f$ in place of f , if necessary.

We will use Lemma 8.1. We take $\lambda = 1 - 16^{-2/\alpha}$. This value of λ is chosen so that 8.3.9 can't hold. Indeed, if $\beta < 1-\lambda$, then

$$||f(x+\beta t_o) + f(x-\beta t_o) - 2f(x)|| \leq M(\beta t_o)^\alpha \leq 16^{-2} Mt_o^\alpha$$

$$\leq \frac{9}{20}(1 - 4\cdot 16^{-2/\alpha}) Mt_o^\alpha$$

$$\leq \frac{9}{20}(\lambda - 2\beta - \beta^2) Mt_o^\alpha \quad .$$

This means that 8.3.8 is valid. Therefore, there is a point $y \in [x_o-\lambda t_o, x_o+\lambda t_o]$ and a quadratic $Q^*(t) = -\frac{9}{20}(1-\lambda)Mt_o^{\alpha-2}(t-y)^2 + \ell(t)$ with ℓ a linear function and

$$Q^*(t) \geq f(t) \qquad t \in [x_o-t_o , x_o+t_o]$$

$$Q^*(y) = f(y) \tag{8.3.14}$$

Now, choose n so that

$$t_n \leq (1-\lambda)\ t_o \leq t_{n-1} \tag{8.3.15}$$

Then $Q^* \geq f$ on $[y-t_n , y+t_n]$ and so

$$f(y+t_n) + f(y-t_n) - 2f(y) \leq Q(y+t_n) + Q(y-t_n) - 2Q(y)$$

$$= -\frac{9}{10}(1-\lambda)Mt_o^{\alpha-2}\ t_n^2$$

$$\leq -\frac{9}{10}(1-\lambda)M(1-\lambda)^{2-\alpha}\ t_{n-1}^{\alpha-2}\ t_n^2$$

$$\leq -\frac{9}{10}(1-\lambda)^{3-\alpha}M\ C^{\alpha-2}\ t_n^{\alpha-2}\ t_n^2$$

$$= -B_\alpha t_n^\alpha$$

where B_α is a positive constant. Here, in the second to last inequality we used (8.3.15) and in the last inequality we used (8.3.13).

We have just shown that

$$t_n^{-\alpha} \, |f(y+t_n) + f(y-t_n) - 2f(y)| \geq B_\alpha M$$

and so

$$||f||^*_{\alpha,1} \leq B_\alpha^{-1} \, ||f||^*_{\alpha,2}$$

whenever $||f||^*_{\alpha,1}$ is finite. The general case is now easy to handle. If $||f||^*_{\alpha,2}$ is finite then mollifying f by convoluting it with say the Fejèr kernels F_m we have

$$||f*F_m||^*_{\alpha,2} \leq ||f||^*_{\alpha,2} \; .$$

Since $||f*F_m||^*_{\alpha,1}$ is finite for each $m \in N$, we have

$$||f*F_m||^*_{\alpha,1} \leq B_\alpha^{-1} \, ||f*F_m||^*_{\alpha,2} \leq B_\alpha^{-1} \, ||f||^*_{\alpha,2} \quad .$$

Taking a limit as $m \to \infty$ establishes that $||f||^*_{\alpha,1}$ is finite, as desired.

The above argument together with Section 8.1 gives the following theorem.

Theorem 8.3. Let $t_n \downarrow 0$ and $0 < \alpha < 2$. A necessary and sufficient condition that for each $f \in C[-a,a]$

$$||\Delta^2_{t_n}(f)|| \; [-a+t_n, a-t_n] = O(t_n^\alpha) \qquad (n\to\infty)$$

be equivalent to

$$||\Delta_t^2(f)|| \; [-a+t,a-t] = O(t^{\alpha}) \qquad (t\to 0)$$

is that

$$\frac{t_n}{t_{n+1}} \le M < +\infty \quad .$$

Remark. The proof above was for $C^*[-\pi,\pi]$. The general case is proved in Theorem 8.4.

We now proceed to develop the general inverse theorem. We define $||f||^*_{\alpha,1}$ as in (8.1.4) with $||\cdot||$ taken on $[-a,a]$ and now give a more general definition for $||\cdot||^*_{\alpha,2}$ as

$$||f||^*_{\alpha,2} = \sup_n \phi_n^{-\alpha} \; ||L_n(f) - f|| \quad .$$

The following lemma shows that $||\cdot||^*_{\alpha,2}$ dominates a suitable multiple of $||\cdot||^*_{\alpha,1}$, which is the crucial step.

Lemma 8.2. *Let* (L_n) *be given by* (8.3.1) *with* (ϕ_n) *defined by* (8.3.2). *Suppose* (ϕ_n) *converges to* 0 , *satisfies* (8.3.5) *and also there is a constant* $C > 0$ *such that*

$$C\phi_n \le \phi_{n+1} \le C^{-1}\cdot \phi_n \qquad n \in N \tag{8.3.16}$$

If $0 < \alpha \le 2$, *then there is a constant* $C_\alpha > 0$ *such that for each* $f \in C[-2a,2a]$ *with* $||f||^*_{\alpha,1}$ *finite, we have*

$$||f||^*_{\alpha,1} \le C_\alpha \; ||f||^*_{\alpha,2} \quad .$$

Proof. There will be a lot of constants appearing in the proof and to make it clear that none of these depend on f , we will prescribe them in the beginning. This hides to some extent why these values are chosen.

First choose $0 < \lambda < 1$, sufficiently close to 1 so that

$$(1-\lambda)^{\alpha} < \frac{9}{20}(\lambda - 2(1-\lambda) - (1-\lambda)^2) \qquad (8.3.17).$$

Fix λ, and choose $\varepsilon > 0$, sufficiently small so that

$$-\frac{9}{20}(1-\lambda)^{3-\alpha}(1-\varepsilon)C^{2-\alpha} + \frac{\varepsilon}{2} < 0 \qquad (8.3.18)$$

where C is the constant in (8.3.16). Now, fix ε, and choose $A \geq a\,\phi_1^{-1}$ so that (8.3.5) holds.

Let f be any function in $C[-2a,2a]$ with $||f||^*_{\alpha,1} = M < +\infty$ and choose $x_0, t_0 \in [-a,a]$ such that

$$f(x_0+t_0) + f(x_0-t_0) - 2f(x_0) \leq -\frac{9}{10} Mt_0^{\alpha} = -\frac{9}{10} Mt_0^{\alpha-2}\, t_0^2 .$$

Here to get the negative sign, we may have to work with $-f$ in place of f.

Lemma 8.1 applies and we want to see that because of our choice of λ, (8.3.9) can't hold. If $0 < \beta < 1-\lambda$, then because of (8.3.17), we have

$$\begin{aligned} |f(y+\beta t_0) + f(y-\beta t_0) - 2f(y)| &\leq M\beta^{\alpha}\, t_0^{\alpha} \leq M(1-\lambda)^{\alpha}\, t_0^{\alpha} \\ &< \frac{9}{20}(\lambda-2(1-\lambda) - (1-\lambda)^2)Mt_0^{\alpha} \\ &< \frac{9}{20}(\lambda-2\beta-\beta^2)\, Mt_0^{\alpha-2}\, t_0^2 . \end{aligned}$$

Hence (8.3.9) can't hold.

Therefore, (8.3.8) is valid. Let y be in $[x_0-\lambda t_0 , x_0+\lambda t_0]$ and $Q^*(t) = -\frac{9}{20}(1-\lambda)Mt_0^{\alpha-2}(t-y)^2 + \ell(t)$ with $Q^*(t) \geq f(t)$, $t \in [x_0-t_0, x_0+t_0]$

and $Q^*(y) = f(y)$. Here, ℓ is a linear function.

Let m be any integer, such that

$$A\phi_m < (1-\lambda)t_o \leq A\phi_{m-1} \qquad (8.3.19).$$

Such an integer exists since $\phi_n \to 0$ and we have imposed the restriction $A \geq a\phi_1^{-1}$. We then have $Q^*(t) \geq f(t)$ on $[y-A\phi_m, y+A\phi_m]$, so that

$$I_1 = \int_{-A\phi_m}^{A\phi_m} \frac{1}{2}(f(y+t) + f(y-t) - 2f(y))d\mu_m(t)$$

$$\leq \frac{1}{2}\int_{-A\phi_m}^{A\phi_m} [Q^*(y+t) + Q^*(y-t) - 2Q^*(y)]\, d\mu_m(t)$$

$$= -\frac{9}{20} Mt_o^{\alpha-2}(1-\lambda)\int_{-A\phi_m}^{A\phi_m} t^2\, d\mu_m(t)\, dt$$

$$\leq -\frac{9}{20} Mt_o^{\alpha-2}(1-\lambda)(1-\varepsilon)\phi_m^2 \quad .$$

Where, for the last inequality we used (8.3.5) and the definition of ϕ_m .

Since $(1-\lambda)\, t_o \leq A\phi_{m-1}$, $t_o^{\alpha-2} \geq \left(\frac{A\phi_{m-1}}{1-\lambda}\right)^{\alpha-2}$ and so

$$I_1 \leq -\frac{9}{20} M(1-\lambda)^{3-\alpha}(1-\varepsilon)(A\phi_{m-1})^{\alpha-2}\, \phi_m^2 \leq -\frac{9}{20} M(1-\lambda)^{3-\alpha}(1-\varepsilon)A^{\alpha-2}C^{2-\alpha}\, \phi_m^{\alpha} \quad (8.3.20)$$

where, we have used (8.3.16).

Now, we will estimate the integral outside $[-A\phi_m, A\phi_m]$. We have

$$|I_2| = \left|\int_{A\phi_m}^{a} \frac{1}{2}(f(y+t) + f(y-t) - 2f(y))d\mu_m(t)\right| \leq \frac{M}{2}\int_{A\phi_m}^{a} |t|^{\alpha}\, d\mu_m(t)$$

$$\leq \frac{M}{2}(A\phi_m)^{\alpha-2}\int_{A\phi_m}^{a} t^2\, d\mu_m(t) \leq \frac{M}{4}(A\phi_m)^{\alpha-2}\, \varepsilon\, \phi_m^2 \qquad (8.3.21)$$

where the last inequality is because of (8.3.5). Of course, if I_3 denotes the integral over $[-a,-A\phi_m]$ then

$$|I_3| \leq \frac{M}{4} (A\phi_m)^{\alpha-2} \varepsilon \phi_m^2 \qquad (8.3.22).$$

Combining (8.3.20), (8.3.21) and (8.3.22), we see that

$$L_m(f,y) - f(y) = I_1 + I_2 + I_3 \leq I_1 + |I_2| + |I_3|$$

$$\leq [-\frac{9}{20} (1-\lambda)^{3-\alpha}(1-\varepsilon)A^{\alpha-2} C^{2-\alpha} + \frac{\varepsilon}{2} A^{\alpha-2}]M\phi_m^\alpha \quad .$$

Using (8.3.18) we see that the term in the brackets is a negative constant, $-B_\alpha$, with $B_\alpha > 0$.

Thus, we have shown that

$$L_m(f,y) - f(y) \leq - B_\alpha M\phi_m^\alpha \qquad (8.3.23).$$

Finally, (8.3.23) implies that

$$B_\alpha M \leq \phi_m^{-\alpha} |L_m(f,y) - f(y)| \leq \phi_m^{-\alpha} ||L_m(f) - f||$$

which proves the lemma with $C_\alpha = B_\alpha^{-1}$.

<u>Remarks</u>. If $0 < \delta < \frac{a}{2}$, then because of (8.3.5)

$$\int_{a-\delta}^{a} d\mu_n(t) = o(\phi_n^2) \quad .$$

Let L_n^δ denote the operators

$$L_n^\delta(f,x) = \int_{-a+\delta}^{a-\delta} \frac{1}{2}(f(x+t) + f(x-t))\, C_n d\mu_n(t)$$

with C_n chosen so that $\int_{-a+\delta}^{a-\delta} C_n\, d\mu_n(t) = 1$. It follows that $C_n = 1 + o(\phi_n^2)$. Now, (L_n^δ) satisfies properties (8.3.5) and (8.3.16) and hence Lemma 8.2 applies. The constant C_α appearing in Lemma 8.2 depends on δ, however, we can choose a universal C_α which will work for all $0 < \delta < \frac{a}{2}$. This is easily checked by noting the dependence of B_α on δ.

Theorem 8.4. Let (L_n) be given by (8.3.1) with (ϕ_n) defined by (8.3.2) and satisfying (8.3.5) and (8.3.16). If $0 < \alpha \leq 2$, then $f \in \mathrm{Lip}^*\alpha$, if and only if $||f - L_n(f)|| = O(\phi_n^\alpha)$.

Proof. The direct part of the theorem follows from our estimate (8.3.3). For the inverse direction, we use the mollifier technique. In the periodic case, there is no difficulty, but in the general, we must be careful at the end points of the interval.

Suppose $f \in C[-2a,2a]$, with $||f||_{\alpha,2}^* = M < +\infty$. If $0 < \delta < \frac{a}{4}$, we let

$$f_\delta(x) = \frac{1}{4\delta^2} \int_{-\delta}^{\delta} \int_{-\delta}^{\delta} f(x+u+v)\,du\,dv \quad .$$

Then, f_δ is twice continuously differentiable and hence $||f_\delta||_{\alpha,1}^*$ is finite. Also, by interchanging integrals we see that

$$|f_\delta(x) - L_n(f_\delta,x)| \leq ||f - L_n(f)|| \leq M\phi_n^\alpha \quad \text{for} \quad |x| \leq a-2\delta \qquad (8.3.24).$$

From (8.3.5), it follows that

$$\int_{a-2\delta}^{a} d\mu_n(t) = o(\phi_n^2)$$

with the "o" uniform in δ, $0 < \delta < \frac{a}{4}$. This means that because of (8.3.24), we can find M' such that

$$\left|\left|\int_{-a+2\delta}^{a-2\delta} \frac{1}{2}(f_\delta(x+t) + f_\delta(x-t) - 2f_\delta(x))C_n(\delta)\, d\mu_n(t)\right|\right|[[-a+2\delta,a-2\delta] \leq M'\phi_n^\alpha \quad (8.3.25)$$

for all n, and $0 < \delta < \frac{a}{4}$. The $C_n(\delta)$ is the normalizing constant for $[-a+2\delta,a-2\delta]$. As we have remarked after Lemma 8.2, we can choose C_α in Lemma 8.2 independent of δ, and so from (8.3.25) we find

$$||f_\delta||^*_{\alpha,1}[-a+2\delta,a-2\delta] \leq C_\alpha M' \quad (8.3.26).$$

$[-a+2\delta,a-2\delta]$ is used to indicate that $||\cdot||^*_{\alpha,1}$ is defined by (8.1.4) over that interval.

Taking a limit as $\delta \to 0$ gives that for each $x \in (-a,a)$, $0 < t < a$

$$\left|\frac{f(x+t) + f(x-t) - 2f(x)}{t^\alpha}\right| \leq C_\alpha M'$$

and the theorem is proved.

<u>8.4. Examples.</u> In many cases, the verification of condition (8.3.5) can be done by calculating the fourth moments of $(d\mu_n)$. We incorporate this in the following corollary to Theorem 8.4.

<u>Corollary 8.2.</u> *Let* (L_n) *be given by* (8.3.1) *with* (ϕ_n) *defined by* (8.3.2) *and satisfying* (8.3.16). *If*

$$\int_{-a}^{a} t^4\, d\mu_n(t) = O(\phi_n^4) \quad (8.4.1).$$

Then for each $0 < \alpha \leq 2$, $f \in Lip^*\alpha$, *if and only if*

$$||f - L_n(f)|| = O(\phi_n^\alpha) \quad .$$

Remark. (8.4.1) can be replaced by the weaker assumption $\int_{-a}^{a} |t|^\gamma d\mu_n(t) = O(\phi_n^\gamma)$, for some $\gamma > 2$. The proof is the same. However, in applications it is usually simplest to work with the fourth moments.

Proof. The only point which has to be verified is that (8.3.5) holds. For $A > 0$,

$$\int_{A\phi_n}^{a} t^2 \, d\mu_n(t) \le (A\phi_n)^{-2} \int_{A\phi_n}^{a} t^4 \, d\mu_n(t)$$

$$\le A^{-2} \phi_n^{-2} \int_{-a}^{a} t^4 \, d\mu_n(t) \le C \, A^{-2} \, \phi_n^2$$

where C is a constant independent of A and n . Therefore, given $\varepsilon > 0$ by choosing A sufficiently large, we see that (8.3.5) is valid.

For an example, we consider the Landau operators ℓ_n introduced in Section 2.5. If $0 < \delta < \frac{1}{2}$, we write

$$\ell_n(f,x) = c_n \int_{-\frac{1}{2}}^{\frac{1}{2}} f(t)(1-(t-x)^2)^n \, dt = c_n \int_{-\frac{1}{2}-x}^{\frac{1}{2}-x} f(x+t)(1-t^2)^n \, dt$$

$$= c_n \int_{-\frac{1}{2}-\delta}^{\frac{1}{2}-\delta} \frac{1}{2}(f(x+t) + f(x-t))(1-t^2)^n \, dt + R_n(f,x)$$

$$= \ell_n^* (f,x) + R_n(f,x) \quad .$$

For any $f \in C[-\frac{1}{2} , \frac{1}{2}]$,

$$||R_n(f)||[-\delta,\delta] \le ||f||\ 2c_n \int_{\frac{1}{2}-\delta}^{1} (1-t^2)^n\, dt$$

$$\le 2||f||\ c_n(\tfrac{1}{2} - \delta)^{-4} \int_{\frac{1}{2}-\delta}^{1} t^4(1-t^2)^n\, dt \le C||f||\ n^{-2} \ .$$

The last inequality was shown in (2.5.3). Here $||\cdot||$ is taken on $[-\frac{1}{2}, \frac{1}{2}]$.

Thus, if $f \in C[-\frac{1}{2}, \frac{1}{2}]$ and $0 < \alpha \le 2$, then

$$||f - \ell_n(f)||[-\delta,\delta] = O(n^{-\alpha/2}) \tag{8.4.2}$$

if and only if

$$||f - \ell_n^*(f)||[-\delta,\delta] = O(n^{-\alpha/2}) \tag{8.4.3}.$$

<u>Theorem 8.5</u>. <u>Let</u> (ℓ_n) <u>be the sequence of Landau operators</u>. <u>If</u> $0 < \delta < \frac{1}{2}$ <u>and</u> $0 < \alpha \le 2$, <u>then for</u> $f \in C[-\frac{1}{2}, \frac{1}{2}]$

$$||f - \ell_n(f)||[-\delta,\delta] = O(n^{-\alpha/2})$$

<u>if and only if</u> $f \in \mathrm{Lip}^*\alpha$ <u>on</u> $[-\delta,\delta]$. <u>That is</u>

$$||\Delta_t^2(f,x)||[-\delta,\delta] = O(t^\alpha) \ .$$

<u>Remark</u>. The inverse portion of this theorem can also be deduced by using the methods of Section 8.2.

<u>Proof</u>. Because (8.4.2) is equivalent to (8.4.3), we need only show that the operators (ℓ_n^*) satisfy the hypothesis of Corollary 8.2. The sequence (ϕ_n^2) for ℓ_n^* is asymptotically equivalent to (n^{-1}) because of our estimates in 2.5.3. We have also shown in that section that the fourth moments for (ℓ_n) are $O(n^{-2})$

from which it follows that (8.4.1) of Corollary 8.2 is valid.

8.5. Inverse Theorems for the Bernstein Operators. The techniques developed in Section 8.3 are also useful in cases in which the operators are not of convolution type. We will not give a general formulation of the technique (see 8.6.3) in this case but choose instead to illustrate the ideas involved in the special, but important, example of Bernstein operators. Inverse theorems for the Bernstein operators were first given by H. Berens and G.G. Lorentz[1] in the following theorem.

Theorem 8.6. *Let* $\phi_n(x) = (x(1-x)n^{-1})^{1/2}$, $n \in N$. *A function* $f \in C[0,1]$ *is in* $\mathrm{Lip}^*\alpha$ *on* $[0,1]$ *if and only if there is a constant* $C > 0$ *such that*

$$|f(x) - B_n(f,x)| \leq C(\phi_n(x))^{\alpha} \qquad 0 \leq x \leq 1 \quad , \quad n \in N \qquad (8.5.1).$$

Recall that $\phi_n^2(x) = B_n((t-x)^2,x)$ (Section 2.7). Regarding the direct estimates, these are already known to us via Section 2.7 in all cases except $\alpha = 1$. The reason is that when $0 < \alpha < 1$, then $\mathrm{Lip}^*\alpha = \mathrm{Lip}\ \alpha$ and when $1 < \alpha \leq 2$ then $\mathrm{Lip}^*\alpha = W^{(1)}(\alpha-1)$. The case $\alpha = 1$ (as well as the general case) is shown in the same way we have proven the general direct Theorem 8.1. Recall, that Theorem 8.1 was shown only for the periodic case. However, as we have remarked, there is an analogous result for $C[a,b]$. The proof is the same as Theorem 8.1 but there is a slight technical difficulty due to the lack of periodicity, which we should point out. In the definition of f_ϵ as given in Theorem 8.1, it is first necessary to extend f to the whole line in such a way that the new function $\overline{f}$ satisfies

$$\omega_2(\overline{f},\delta) \leq 5\omega_2(f,\delta) \qquad \delta > 0 \quad .$$

We have noted in (1.3.7) that such a construction is possible. It is then necessary to introduce the multiplicative constant 5 on the right hand sides of (8.2.3),

(8.2.4), and (8.2.5), but this does not effect the argument.

Regarding the inverse direction, H. Berens and G.G. Lorentz[1] have given an elementary proof in the case that $0 < \alpha < 1$ by using inequalities for $(B_n(f))'$. The proof for the general case is more difficult and depends on delicate estimates for $(B_n(f))''$. Our approach to the inverse theorem is based on the idea of Section 8.3. We begin with the following estimates for the concentration of "mass" of B_n.

Lemma 8.3. Let $g(y,c,t)$ be the function which is 0 for $|t-y| \leq c$ and 1 otherwise. If $A > 0$ and $y(1-y) \geq n^{-1}$, $y \in [0,1]$, then

$$B_n(g(y,A\phi_n(y),t),y) \leq 4A^{-4} \tag{8.5.2}$$

$$B_n(|t-y|g(y,A\phi_n(y),t),y) \leq 4A^{-3}\,\phi_n(y) \tag{8.5.3}$$

and

$$B_n((t-y)^2 g(y,A\phi_n(y),t),y) \leq 4A^{-2}\,\phi_n^2(y) \tag{8.5.4}$$

Proof. The following identity follows from the recurrence formula (2.7.3)

$$B_n((t-y)^4,y) = (3-2n^{-1})\phi_n^4(y) + (1-2y)^2\, n^{-2}\,\phi_n^2(y) \tag{8.5.5}$$

This shows that

$$B_n((t-y)^4,y) \leq 4\phi_n^4(y) \tag{8.5.6}$$

when $y(1-y) \geq n^{-1}$.

To establish (8.5.2), we observe that $A^{-4}\ \phi_n^{-4}(y)(t-y)^4 \geq g(y,A\phi_n(y),t)$ and so

$$B_n(g(y,A\phi_n(y),t),y) \leq A^{-4}\ \phi_n^{-4}(y)B_n((t-y)^4,y) \leq 4A^{-4}$$

where in the last inequality we have used (8.5.6). The other two inequalities are verified in the same fashion.

Lemma 8.4. Let $0 < \alpha \leq 2$ and $C = 20^{-2/\alpha}$. If $f \in \mathrm{Lip}^* \frac{\alpha}{2}$ on $[0,1]$ but not in $\mathrm{Lip}^*\alpha$ on $[0,1]$, then for each $M > 0$, there are points $x_o \in [0,1]$, $t_o \in [x_o,1-x_o]$ and an $M_o \geq M$ such that

$$|f(x_o+t_o) + f(x_o-t_o) - 2f(x_o)| = M_o\ t_o^{\alpha} \qquad (8.5.7)$$

and if $\beta \leq C_\alpha$, $x \in [\beta t_o,(1-\beta)t_o]$

$$|f(x+\beta t_o) + f(x-\beta t_o) - 2f(x)| \leq 10^{-1}\ M_o\ t_o^{\alpha} \qquad (8.5.8).$$

Proof. Certainly, there are points x_1,t_1 such that if we let $x_o = x_1$, $t_o = t_1$ then (8.5.7) will be satisfied with an appropriate $M_o = M_1$. If for this choice of x_o,t_o we do not have (8.5.8) satisfied then we construct a new pair of points $x_2,\beta_2 t_1$ where $\beta_2 \leq C_\alpha$ and

$$|f(x_2+\beta_2 t_1) + f(x_2-\beta_2 t_1) - 2f(x_2)| \geq 10^{-1}\ M_1 t_1^{\alpha} \geq M(\beta_2 t_1)^{\alpha} \qquad (8.5.9).$$

The inequality (8.5.9) shows that (8.5.7) will hold for $x_o = x_2$, $t_o = \beta_2 t_1$ and of course an appropriate choice for M_o. If this choice for x_o and t_o don't satisfy (8.5.8) then we can repeat our argument to construct points $x_3,\beta_3\beta_2 t_1$.

We want to show that if we continue this construction at some point we must have (8.5.8) satisfied. If $x_n, \beta_n \cdots \beta_2 t_1$ are the points constructed at the nth step, then

$$|f(x_n+\beta_n \cdots \beta_2 t_1) + f(x_n-\beta_n \cdots \beta_2 t_1) - 2f(x_n)| \geq 10^{-n+1} M_1 t_1^{\alpha} \quad (8.5.10).$$

Since $f \in \mathrm{Lip}^* \frac{\alpha}{2}$, for some constant $C > 0$,

$$(f(x_n+\beta_n \cdots \beta_2 t_1) + f(x_n-\beta_n \cdots \beta_2 t_1) - 2f(x_n)| \leq C(\beta_2 \cdots \beta_n)^{\frac{\alpha}{2}} t_1^{\frac{\alpha}{2}}$$

$$\leq C\, C_\alpha^{\frac{1}{2}(n-1)\alpha}\, t_1^{\frac{1}{2}\alpha} \leq 20^{-n+1}\, C\, t_1^{\frac{1}{2}\alpha} \quad (8.5.11).$$

Therefore (8.5.11) will contradict (8.5.10) for n suitably large and the lemma is proved.

Proof of Theorem 8.6. We want to prove the inverse direction, namely, (8.5.1) implies $f \in \mathrm{Lip}^* \alpha$. We have already discussed the other direction. Suppose then, that $f \in C[0,1]$ and satisfies (8.5.1). It is enough to consider the case where the constant C in (8.5.1) is equal to 1 which we do. Since $B_n(f)$ is an algebraic polynomial of degree $\leq n$, Theorem 1.7 shows that $f \in \mathrm{Lip}^* \frac{\alpha}{2}$. We suppose $f \notin \mathrm{Lip}^* \alpha$ and work for a contradiction.

Let $C_\alpha = 20^{-2/\alpha}$ as in Lemma 8.4 and let $M = 10^5 C_\alpha^{-9}$ (there will be large overkill in our choice of constants). We let x_0, t_0, M_0 be as in Lemma 8.4. By working with $-f$ in place of f, if necessary, we can suppose that

$$f(x_0+t_0) + f(x_0-t_0) - 2f(x_0) = -M_0\, t_0^{\alpha}$$

with $M_0 \geq M$.

We can also require that

$$\omega_2(f,t_o) = M_o\ t_o^{\alpha} \qquad (8.5.12).$$

To see this, we know that there is some $h \le t_o$ and x_o' such that

$$|f(x_o'+h) + f(x_o'-h) - 2f(x_o')| = \omega_2(f,h) = \omega_2(f,t_o) \ge M_o\ t_o^{\alpha} \quad .$$

Therefore, writing $\omega_2(f,t_o) = M_o'\ h^{\alpha}$, we have $M_o' \ge M_o$. Lemma 8.4 will now be satisfied with the new choices of M_o' for M_o, and h for t_o. These choices will also satisfy (8.5.12).

Now let $g = f-\ell$ where ℓ is the linear function which interpolates f at x_o-t_o and x_o+t_o. The function g has the same symmetric second divided differences as f.

We will use Lemma 8.1 with $\lambda = 1-C_{\alpha}$. If $\beta < 1-\lambda = C_{\alpha}$, and $x \in [0,1]$, then

$$|g(x+\beta t_o) + g(x-\beta t_o) - 2g(x)| \le 10^{-1}\ M_o t_o^{\alpha} < \tfrac{1}{2}(\lambda-2\beta-\beta^2)\ M_o\ t_o^{\alpha}$$

because of (8.5.8) and $C_{\alpha} \le \frac{1}{8}$. This shows that (8.3.9) of Lemma 8.1 can't hold and therefore (8.3.8) is valid. In otherwords, there is a quadratic $Q(t) = -\frac{1}{2}\ C_{\alpha}M_o t_o^{\alpha-2}(t-x_o)^2 + C$, with C a constant, and

$$Q(t) \ge g(t) \quad , \quad t \in [x_o-t_o, x_o+t_o] \qquad (8.5.13)$$

and

$$Q(y) = g(y) \qquad (8.5.14)$$

where $y \in [x_o-(1-C_\alpha)t_o\ ,\ x_o+(1-C_\alpha)t_o]$. Here the linear term in Q does not appear since g vanishes at x_o-t_o and x_o+t_o (see proof of Lemma 8.1).

If we denote by h the function which is 0 on $[x_o-t_o\ ,\ x_o+t_o]$ and 1 otherwise then $g \leq Q + (g-Q)h$ on $[0,1]$. Let $A = 100\ C_\alpha^{-3}$ and choose m such that

$$A\phi_m(y) \leq C_\alpha t_o \leq A\phi_{m-1}(y) \tag{8.5.15}$$

Such an m exists because

$$A\phi_1(y) = A(y(1-y))^{1/2} \geq AC_\alpha t_o \geq C_\alpha\ t_o\ .$$

We have

$$\begin{aligned} B_m(g,y) - g(y) &\leq B_m(Q,y) - Q(y) + B_m(|g(t) - g(y)|h(t),y) + B_m(|Q(t)-Q(y)|h(t),y) \\ &= I_1 + I_2 + I_3 \end{aligned} \tag{8.5.16}$$

Since B_m preserves linear functions we have

$$I_1 = -\frac{1}{2}\, C_\alpha\, M_o\, t_o^{\alpha-2}\, \phi_m^2(y) \tag{8.5.17}$$

We will now show that $I_2, I_3 \leq \frac{1}{5}\, C_\alpha\, M_o t_o^{\alpha-2}\ \phi_m^2(y)$.

First for I_3 , we see that

$$\begin{aligned} |Q(t)-Q(y)| &= \left|\frac{1}{2}\, C_\alpha\, M_o t_o^{\alpha-2}(t-y)^2 + C_\alpha\, M_o t_o^{\alpha-2}(y-x_o)(t-y)\right| \\ &\leq \frac{1}{2}\, C_\alpha\, M_o t_o^{\alpha-2}\ [\ (t-y)^2 + 2t_o|t-y|] \end{aligned} \tag{8.5.18}$$

We want to use Lemma 8.3, so we must show that $y(1-y) \geq m^{-1}$. From (8.5.15), we find

$$A^2 \frac{y(1-y)}{m} \leq C_\alpha^2 t_o^2 .$$

Since $y \in [x_o-(1-C_\alpha)t_o , x_o+(1-C_\alpha)t_o]$, it follows that

$$y(1-y) \geq \frac{1}{2} C_\alpha t_o .$$

These two inequalities combine to show that

$$\frac{A^2}{m} \leq 2 C_\alpha t_o \leq 4y(1-y) .$$

Since $A^2 > 4$, we do have $y(1-y) \geq m^{-1}$.

Recall now, that the function h vanishes on $[x_o-t_o , x_o+t_o]$ and hence on $[y - A\phi_m(y) , y + A\phi_m(y)]$. So, if we apply B_m to both sides of (8.5.18) and use Lemma 8.3, we see

$$\begin{aligned} I_3 = B_m(|Q(t)-Q(y)|h(t),y) &\leq 2C_\alpha M_o t_o^{\alpha-2}[A^{-2} \phi_m^2(y) + 2t_o A^{-1} \phi_m(y)] \\ &\leq 2C_\alpha M_o t_o^{\alpha-2} A^{-2} \phi_m^2(y) (1 + 4C_\alpha^{-1}) \\ &\leq \frac{1}{5} C_\alpha M_o t_o^{\alpha-2} \phi_m^2(y) \end{aligned} \tag{8.5.19}$$

where in the second inequality we used the fact that $A^{-1}t_o \leq C_\alpha^{-1} \phi_{m-1}(y) \leq 2C_\alpha^{-1} \phi_m(y)$. The last inequality in (8.5.19) comes from our choice of A.

We now estimate I_2.

We have noted in the beginning of this section that because of (8.5.12) there is a twice continuously differentiable function g_{t_0} such that

$$||g-g_{t_0}|| \leq 5\, M_0\, t_0^{\alpha} \qquad (8.5.20)$$

$$||g''_{t_0}|| \leq 5\, M_0 t_0^{\alpha-2} \qquad (8.5.21).$$

Because of (8.5.20) and the fact that g vanishes at $x_0 \pm t_0$, we see that

$$|g_{t_0}(x_0 \pm t_0)| \leq 5\, M_0\, t_0^{\alpha} .$$

Therefore, using the mean value theorem, there is a point $\xi \in [x_0 - t_0\, ,\, x_0 + t_0]$ with

$$|g'_{t_0}(\xi)| \leq 5\, M_0\, t_0^{\alpha-1} \qquad (8.5.22).$$

Using (8.5.21) with (8.5.22), we find

$$\begin{aligned}|g'_{t_0}(y)| &\leq |g'_{t_0}(\xi)| + |g'_{t_0}(y) - g'_{t_0}(\xi)| \\ &\leq 5\, M_0\, t_0^{\alpha-1} + 5\, M_0\, t_0^{\alpha-2}\, |\xi-y| \\ &\leq 15\, M_0\, t_0^{\alpha-1} \qquad (8.5.23).\end{aligned}$$

This means we can estimate

$$\begin{aligned}|g(t)-g(y)| \leq\, & |g_{t_0}(t) - g_{t_0}(y) - g'_{t_0}(y)(t-y)| + |g(t) - g_{t_0}(t)| \\ & + |g_{t_0}(y) - g(y)| + |g'_{t_0}(y)|\; |t-y|\end{aligned}$$

$$\leq 5\, M_o\, t_o^{\alpha-2}(t-y)^2 + 10\, M_o\, t_o^{\alpha} + 15\, M_o\, t_o^{\alpha-1}\, |t-y| \qquad (8.5.24).$$

Where, we used (8.5.21) in estimating the first term, (8.5.20) in estimating the middle two terms, and (8.5.23) in estimating the last term. Now applying B_m to both sides of (8.5.24) and using Lemma 8.3 we find

$$I_2 \leq 4\, M_o\, t_o^{\alpha-2} A^{-2}\left(5\phi_m^2(y) + 10\, A^{-2}\, t_o^2 + 15\, A^{-1}\, t_o \phi_m(y)\right)$$

$$\leq 4\, C_\alpha\, M_o\, t_o^{\alpha-2}\, \phi_m^2(y) A^{-2}\, (5\, C_\alpha^{-1} + 40\, C_\alpha^{-3} + 30\, C_\alpha^{-2})$$

$$\leq \frac{1}{5}\, C_\alpha\, M_o\, t_o^{\alpha-2}\, \phi_m^2(y) \qquad (8.5.25)$$

where in the second inequality we used that $A^{-1}\, t_o \leq 2\, C_\alpha^{-1}\, \phi_m(y)$. The last inequality comes from the value of A.

Finally, combining (8.5.17), (8.5.19) and (8.5.25) shows that

$$B_m(f,y) - f(y) = B_m(g,y) - g(y) \leq -\, 10^{-1}\, C_\alpha\, M_o\, t_o^{\alpha-2}\, \phi_m^2(y)$$

because B_m preserves linear functions. Since $t_o \leq AC_\alpha^{-1}\, \phi_{m-1}(y) \leq 2AC_\alpha^{-1}\, \phi_m(y)$ we find

$$|B_m(f,y)-f(y)| \geq \frac{2^{\alpha-2}}{10}\, C_\alpha^{3-\alpha} A^{\alpha-2}\, M_o(\phi_m(y))^\alpha \geq 40^{-1}\, C_\alpha^{3-\alpha} A^{\alpha-2}\, M_o(\phi_m(y))^\alpha \quad (8.5.26).$$

But, since $M_o \geq M$

$$40^{-1}\, C_\alpha^{3-\alpha}\, A^{\alpha-2}\, M_o > 1$$

because of our choices of $M = 10^5\, C_\alpha^{-9}$ and $A = 100\, C_\alpha^{-3}$. This shows that (8.5.26) contradicts (8.5.1) and the proof is complete.

8.6 Notes.

8.6.1 In Theorem 8.3, it was relatively difficult to show that condition (8.1.3) is sufficient to guarantee that whenever $||\Delta_{t_n}^2 f|| = O(t_n^\alpha)$ then $f \in \mathrm{Lip}^*\alpha$. For first divided differences there is a simpler proof of the corresponding result. First, we can always assume that the sequence (t_n) has the property that $\delta^{-1} \leq t_n^{-1} \leq 2^{-1}$, $n \in N$, with $\delta > 2$. This is because we can always extract a subsequence which has this property and then work with the subsequence in place of (t_n) (see the construction in the proof of Theorem 8.2).

Now if $t > 0$, then there exists a subsequence (t_{n_k}) and non-negative integers β_k such that $t = \sum_{k=1}^{\infty} \beta_k t_{n_k}$ with the β_k's uniformly bounded. Such a construction is made inductively by letting n_1 be the smallest integer such that $t_{n_1} \leq t$ and taking β_1 as the largest integer for which $\beta_1 t_{n_1} \leq t$. Having constructed $\beta_1, \cdots, \beta_k$ and $n_1, \cdots, n_k$, we let n_{k+1} be the smallest integer such that $t_{n_{k+1}} \leq t - \sum_{j=1}^{k} \beta_j t_{n_j}$ and β_{k+1} as the largest integer such that $\beta_{k+1} t_{n_{k+1}} \leq \sum_{j=1}^{k} \beta_j t_{n_j}$.

Having made such a construction, we find that if $||f(x+t_n)-f(x)|| = O(t_n^\alpha)$ then

$$||f(x+t)-f(x)|| \leq \sum_{k=1}^{\infty} ||f(x + \sum_{j=1}^{k} \beta_j t_{n_j}) - f(x + \sum_{j=1}^{k-1} \beta_j t_{n_j})||$$

$$\leq \sum_{k=1}^{\infty} \beta_k t_{n_k}^\alpha \leq C t_{n_1}^\alpha \sum_{k=1}^{\infty} 2^{-k\alpha} \leq C' t^\alpha$$

with C' independent of t. Hence, $f \in \mathrm{Lip}\,\alpha$.

The necessity of (8.1.3) for having $||\Delta_{t_n} f|| = O(t_n^{\alpha})$ imply $f \in \text{Lip}\ \alpha$ can be shown in the same way we have argued in Section 8.1 for $\Delta_{t_n}^2$.

<u>8.6.2</u> The techniques developed in Sections 8.3 and 8.5 for proving inverse theorems can not be used in the L_p - spaces. In fact, for general L_p, it has not yet been shown that (8.1.3) is sufficient for $||\Delta_{t_n}^2 f||_p = O(t_n^{\alpha})$, to imply $||\Delta_t^2 f||_p = O(t^{\alpha})$. The necessity of (8.1.3) can be shown as we have argued in Section 8.1. Also, in a forthcoming paper G. Freud will give a constructive proof of the necessity of (8.1.3). Freud also will show that (8.1.3) is sufficient for L_2.

<u>8.6.3</u> The reason we have not put the technique of Section 8.5 in the form of a general theorem is that in order to use these ideas we must have some a priori information about the smoothness of f. For Bernstein polynomials this came from the general inverse theorems for approximation by algebraic polynomials which gave that $f \in \text{Lip}^* \frac{\alpha}{2}$ whenever (8.5.1) holds. In general, such information is not available. It should be possible to modify the technique to circumvent this problem.

REFERENCES

Alexits, G.

1. Sur l'ordre de grandeur de l'approximation d'une fonction par les moyennes de sa serie Fourier, (Hungar.) Mat. Fizikai Lapok, 48(1941), 410-422.

Aljancic, S.

1. Approximation of continuous functions by typical means of their Fourier series, Proc. A.M.S., 12(1961), 681-688.

Amelkovič, V.G.

1. A theorem converse to a theorem of Voronovskaja type, (Russian), Teor. Funkciĭ Funkcional Anal. i. Priložen, 2(1966), 67-74.

Bajšanski, B. and Bojanic, R.

1. A note on approximation by Bernstein polynomials, Bull. A.M.S., 70 (1964), 675-677.

Baskakov, V.A.

1. The degree of approximation of differentiable functions by certain positive operators (Russian), Mat. Sbornik 76(118)(1968), 344-361.

Berens, H.

1. Interpolationsmethoden zur Behandlung von Approximationsprozessen auf Banachräumen, Springer Lecture Notes in Math., Vol. 64.
2. On the saturation theorem for Cesaro means of Fourier series, Acta Math. Acad. Sci. Hung., 21(1970), 95-99.
3. On pointwise approximation of Fourier series by typical means, Tohoku Math. J., 23(1971), 147-153
4. Pointwise saturation of positive operators (to appear).

Berens, H., and Lorentz, G.G.

1. Inverse theorems for Bernstein polynomials, (to appear in Indiana J. of Math. and Mech.).

Bohman, H.

1. On approximation of continuous and of analytic functions, Ark. Mat., 2(1952), 43-56.

Bojanic, R.

1. A note on the degree of approximation to continuous functions, L'Enseignement Math., 15(1969), 43-51.
2. A note on the precision of interpolation by Hermite-Fejér polynomials, (to appear in the Proceedings of the Conference on the Constructive Theory of Functions, Budapest, 1969).

Bojanic, R., and DeVore, R.

1. A proof of Jackson's theorem, Bull. A.M.S., 75(1969), 364-367.

Boman, J.

1. Saturation problems and distribution theory, in: Topics in Approximation Theory by H.S. Shapiro, Springer Lecture Notes in Math., Vol. 187.

Buchwalter, H.

1. Saturation de certains procédés de sommation, Comptes Rendus, Acad. Sci. Paris, 248(1959), 909-912.

Butzer, P.L.

1. Sur le role de la transformation de Fourier dans quelques problemes d'approximation, C.R. Acad. Sci. Paris, 249(1959), 2467-2469.
2. Representation and approximation of functions by general singular integrals, Neder. Akad. Wetensch. Proc. Ser. 63A (Indag. Math. 22) (1960), 1-24.

3. Fourier transform methods in the theory of approximation, Arch. Rat. Mech. Anal. 5 (1960), 390-415.

Butzer, P.L., and Berens, H.

1. Semi-groups of Operators and Approximation, Springer, Berlin, 1967 , 318 pp.

Butzer, P.L., and Görlich, E.

1. Saturationsklassen und asymptotische Eingenschaften trigonometrischer singulärer Integrale, in: Festschrift zur Gedachtnisfeier Karl Weierstrass,1815 - 1965, 339-392.

Butzer, P.L., and Nessel, R.J.

1. Fourier Analysis and Approximation, Vol. 1, Academic Press, N.Y. 1971 , 553 pp.

Butzer, P.L., Nessel, R.J., and Scherer, K.

1. Trigonometric convolution operators with kernels having alternating signs and their degree of convergence, Jber. Deutsch. Math. - Verein. 70(1967), 86-99.

Butzer, P.L., and Scherer, K.

1. Approximationsprozesse und Interpolationsmethoden (Hochschulskripten 826/826a), Bibliograph. Inst., Mannheim, 1968 , 172 pp.

Butzer, P.L., and Stark, E.

1. Wesentliche asymptotische Entwicklunen fur Approximationsmasse trigonometrischer singulärer Integrale, Math. Nachr. 39(1969), 223-237.

Censor, E.

1. Quantatative results for positive linear approximation operators, J. Approx. Th. 4(1971), 442-450.

Curtis, P.C.

1. The degree of approximation by positive convolution operators, Mich. Math. J., 12(1965), 155-160.

Davis, P.J.

1. Interpolation and Approximation, Blaisdell, N.Y., 1963 , 393 pp.

De Leeuw, K.

1. On the degree of approximation by Bernstein polynomials. J. d'Anal. Math., 7(1959), 89-104.

DeVore, R.

1. On Jackson's Theorem, J. Approx. Th., 1(1968), 314-318.
2. Optimal convergence of positive linear operators, (to appear in the Proceedings of the Conference on the Constructive Theory of Functions, Budapest, 1969).
3. Saturation of positive convolution operators, J. Approx. Th. 3(1970), 410-429.
4. On a saturation theorem of Tureckii, Tohoku Math. J., 23(1971), 353-362.
5. On the direct theorem of saturation, Tohoku Math. J., 23(1971), 363-370.
6. A pointwise "o" saturation theorem for positive convolution operators, (to appear in the Proceedings of the Conference on Linear Operators and Approximation, Oberwolfach, 1971).

Ditzian, Z.

1. Convergence of sequences of linear positive operators, remarks and applications, (to appear in J. Approx. Th.).

Dunford, N., and Schwartz, J.T.

1. Linear Operators, Part I, Interscience, New York, 1957, 858 pp.

Erdös, P., and Turan, P.

1. On interpolation III, Annals of Math., 41(1940), 510-553.

Favard, J.

1. Sur l'approximation des fonctions d'une variable reèlle, In: Analyse Harmonique, Coll. Int. Centre Rech. Sci., No. 15, Paris (1949), 97-110.
2. Sur l'approximation dans les éspaces vectoriels, Ann. Mat. Pura Appl. 29(1949), 259-291.

Fejér, L.

1. Über trigonometrische Polynome, J. Reine Angew. Math. 146(1916), 53-82.

Freud, G.

1. Über einen Satz von P. Erdös und P. Turan, Acta. Math. Acad. Sci. Hung. 4(1953), 255-266.

2. On approximation by positive linear methods, I & II, Studia Sci. Math. Hung. 2(1967), 63-66, 3(1968), 365-370.

Görlich, E.

1. Distributional methods in saturation theory, J. Approx. Th. 1(1968) 111-136.
2. Uber optimale approximationsoperatoren, Proc. of International Conference on Constructive Function Theory, Varna, 1970, 187-191.

Görlich, E., and Stark, E.L.

1. A unified approach to three problems on approximation by positive linear operators (to appear in the Proceedings of the Conference on the Constructive Theory of Functions, Budapest, 1969).

2. Über beste Konstanten und asymptotische Entwicklungen positiver Faltungsintegrale und deren Zusammenhang mit dern Saturationsproblem, Jber. Deutsch. Math. - Verein., 72(1970), 18-61.

Haršiladge, F.I.

1. Saturation classes for some summation processes (Russian), Doklady, S.S.S.R., 122(1958), 352-355.

Hoff, C.J.

1. Approximation with kernels of finite oscillation I, Convergence, J. Approx. Th., 3(1970), 213-228.

Ikeno, K., and Suzuki, Y.

1. Some remarks on saturation problems in local approximation, Tohuku Math. J., 20(1968), 214-233.

Jackson, D.

1. The Theory of Approximation, Amer. Math. Soc. Colloquium Publ. Vol. XI, N.Y., 1930 , 178 pp.

Karlin, S., and Studden, W.J.

1. Tchebycheff Systems; With Applications in Analysis and Statistics, Interscience, N.Y., 1966, 586 pp.

Karlin, S., and Ziegler, Z.

1. Iterations of positive approximation operators, J. Approx. Th., 3(1970), 310-339.

Katznelson, Y.

1. An Introduction to Harmonic Analysis, Wiley, New York, 1968 , 264 pp.

Knopp, K.

1. Theory and Application of Infinite Series, Blackie, London, 1949 , 571 pp.

Komleva, E.A.

1. The asymptotic properties of positive summation methods for Fourier series, (Russian), Isv. Vyss. Učebn. Zaved. Mat., 1959, no. 4 (11), 89-93.

Korovkin, P.P.

1. On convergence of linear positive operators in the space of continuous functions (Russian), Doklady, S.S.S.R., 90(1953), 961-964.

2. On the order of the approximation of functions by linear positive operators (Russian), Doklady, S.S.S.R., 114(1957), 1158-1161.

3. An asymptotic property of positive methods of summation of Fourier series and best approximation of functions of class Z_2 by linear positive polynomials operators (Russian), Usephi Mat. Nauk, 13(1958), No. 6(84), 99-103.

4. Linear Operators and Approximation Theory, Hindustan, Delhi, 1960, 222 pp.

5. Best approximation of functions of class Z_2 by certain linear operators, (Russian), Doklady, S.S.S.R., 127 (1959), 513-515.

6. Asymptotic properties of positive methods for summation of Fourier series (Russian), Usephi Mat. Nauk, 15(1960), no. 1 (91), 207-212.

Krylov, V.I.

1. Approximate Calculations of Integrals, MacMillan, New York, 1962, 357 pp.

Lorentz, G.G.

1. Inequalities and the saturation classes of Bernstein polynomials, in: On Approximation Theory, Proceedings of the conference held in Oberwolfach, 1963, 200-207.

2. Approximation of Functions, Holt, New York, 1966, 188 pp.

Lorentz, G.G., and Schumaker, L.

1. Saturation of positive operators, (to appear).

Mamedov, R.G.

1. On the order of the approximation of differentiable functions by linear positive operators, (Russian), Doklady, S.S.S.R., 128(1959), 674-676.

2. On the asymptotic value of the approximation of repeadedly differentiable functions by linear positive operators (Russian), Doklady, S.S.S.R., 146 (1962), 1013-1016.

3. On the order and on the asymptotic value of the approximation of functions by generalized Landau operators, (Russian), Akad. Nauk. Azerbaidžanskŏi, S.S.S.R. Trudy Institute Math. i Mechaniki, 2(10)(1963), 49-65.

Marsden, M.

1. An identity for spline functions with applications to variation diminishing spline approximation. J. Approx. Th., 3(1970), 7-49.

Matsuoka, Y.

1. On the approximation of functions by some singular integrals, Tohoku Math. J., 18(1966), 13-43.

Micchelli, C.C.

1. The saturation class and iterates of the Bernstein polynomials, (to appear).

Mühlbach, G.

1. Operatoren vom Bernsteinchen Typ , J. Approx. Th., 3 (1970), 274-292.

Müller, M.W.

1. Approximation durch lineare Positive Operatoren bei Gemischter Norm Habilationsschrift, Stuttgart, 1970.

Müller, M.W., and Walk, H.

1. Konvergenz - und Güteaussagen für die Approximation durch Folgen linearer positiver Operatoren, (to appear in the Proceedings of the Conference on Constructive Function Theory, Varna, 1970).

Natanson, I.P.

1. Saturation classes in the theory of singular integrals (Russian), Doklady, S.S.S.R., 158(1964), 520-523.

Nessel, R.J.

1. Über Nikolskii-Konstanten von positiven Approximationsverfahren bezuglich Lipschitz-Klassen, Jber. Deutsch. Math., Verein., 73 (1971), 6-47.

Newman, D.J., and Shapiro, H.S.

1. Jackson's theorem in higher dimensions, in: On Approximation Theory, Proceedings of the Conference held in Oberwolfach, 1963, 208-218.

Pesin, I.

1. Classical and Modern Integration Theories, Academic Press, New York, 1951, 195 pp.

Popoviciu, T.

1. Sur l'approximation des fonctions convexes d'ordre superieur, Mathematica (Cluj), 10(1935), 49-54.

2. A supra demonstratici teoremei lui Wierstrass cu ajutoral polinoamelor de interpolare, Academia, Rep. Pop. Romine, Lucrarile Sessiunii generale stintifice din 2-12 iunie 1950(1951), 1664-1667.

Šaškin, Ju. A.

1. Korovkin systems in the space of continuous functions, Isv. Akad. Nauk. S.S.S.R., 26(1966), 495-512, A.M.X. transl. 29(1966), 125-144.

Schnabl, R.

1. Zum globalen Saturationsproblem der Folge der Bernstein-operatoren, Acta. Sci. Math. Hung 31(1970), 351-358.

Schoenberg, I.J.

1. On spline functions in: Inequalities, Proceedings of the Conference held at Wright Patt. A.F.B., Ohio, 1965, 255-291.

Shapiro, H.S.

1. Smoothing and Approximation of Functions, Van Nostrand, N.Y., 1969, 134 pp.

Shisha, O., and Mond, B.

1. The degree of convergence of sequences of linear positive operators, Proc. Nat. Acad. Sci. U.S.A. 60(1968), 1196-1200.

Shisha, O., Sternin, C., and Fekete, M.

1. On the accuracy of approximation to given functions by certain interpolatory polynomials of given degree, Riveon Lematematica, 8 (1954), 59-64.

Stark, E.L.

1. Über die Approximationsmasse spezieller singulärer Integrale, Computing 4 (1969), 153-159.

2. On approximation improvement for trigonometric singular integrals by means of finite oscillation kernels with seperated zeros (to appear in the Proceedings of the Conference on Constructive Function Theory, Varna, 1970).

3. Über trigonometrische singuläre Faltungsintegrale mit Kernen endlicher Oszillation, Dissertation, Aachen, 1970.

4. The complete asymptotic expansion for the measure of approximation of Abel-Poisson's singular integral for Lip 1, (to appear in Mat. Zametki).

Stečkin, S.B.

1. On the approximation of periodic functions by Fejer sums (Russian), Trudy Mat. Inst. V.A. Stecklova, 62(1961), 48-60.

Sunouchi, G.

1. Characterization of certain classes of functions, Tohoku Math. J., (2) 14 (1962), 127-134.

2. Saturation in the local approximation, Tohoku Math. J., 17(1965), 16-28.

3. Direct theorems in the theory of approximation, Acta. Math. Sci. Hung. 20 (1969), 409-420.

Sunouchi, G., and Watari, C.

1. On determination of the class of saturation in the theory of approximation I - Proc., Jap. Acad. 34(1958), 477-481; II - Tohoku Math. J. 11 (1959), 480-488.

Suzuki, Y.

1. Saturation of local approximation by linear positive operators, Tohoku Math. J. 17 (1965), 201-221.

Suzuki, Y., and Watanabe, S.

1. Some remarks on saturation problem in the local approximation II, Tohoku Math. J., 21 (1969), 65-83.

Szegö, G.

1. Orthogonal Polynomials, Amer. Math. Soc. Coll. Publ. Vol. XXIII, N.Y. 1959, 421 pp.

Teljakovskii, S.A.

1. The approximation by Fejér sums of functions which satisfy a Lipschitz condition, (Russian), Ukrain. Math. Z. 21(1969), 334-343.

Timan, A.F.

1. Theory of Approximation of Functions of a Real Variable, New York, Macmillan, New York, 1963, 286 pp.

Tureckiĭ, A.H.

1. Saturation in the space C (Russian), Dokl. S.S.S.R., 126 (1959), No. 6, 1207-1209.

Voronovskaja, E.V.

1. The asymptotic behaviour of the approximation of functions by their Bernstein polynomials (Russian), Doklady, S.S.S.R., 4 (1932), 79-85.

Watnabe, S., and Suzuki, Y.

1. Approximation of functions by generalized Meyer-König Zeller operators, Bull. Yamagata U. (Nat. Science) 7 (1969), 123-128.

Wulbert, D.

1. Convergence of operators and Korovkin's theorem, J. Approx. Th. 1 (1968), 381-390.

Zamansky, M.

1. Classes de saturation de certaines procédés d'approximation des series de Fourier des fonctions continues et applications a quelques problemes d'approximation, Ann. Sci. Ecole Norm Sup. 66 (1949), 19-93.

Ziegler, Z.

1. Linear approximation and generalized convexity, J. Approx. Th., 1 (1968), 420-443.

Zygmund, A.

1. Trigonometric Series, Vol. I/II, Cambridge Press, 1959, 383 pp., 364 pp.

Lecture Notes in Mathematics

Vol. 178: Th. Bröcker und T. tom Dieck, Kobordismentheorie. XVI, 191 Seiten. 1970. DM 18,-

Vol. 179: Seminaire Bourbaki - vol. 1968/69. Exposés 347-363. IV. 295 pages. 1971. DM 22,-

Vol. 180: Séminaire Bourbaki - vol. 1969/70. Exposés 364-381. IV, 310 pages. 1971. DM 22,-

Vol. 181: F. DeMeyer and E. Ingraham, Separable Algebras over Commutative Rings. V, 157 pages. 1971. DM 16,-

Vol. 182: L. D. Baumert. Cyclic Difference Sets. VI, 166 pages. 1971. DM 16,-

Vol. 183: Analytic Theory of Differential Equations. Edited by P. F. Hsieh and A. W. J. Stoddart. VI, 225 pages. 1971. DM 20,-

Vol. 184: Symposium on Several Complex Variables, Park City, Utah, 1970. Edited by R. M. Brooks. V, 234 pages. 1971. DM 20.

Vol. 185: Several Complex Variables II, Maryland 1970. Edited by J. Horváth. III, 287 pages. 1971. DM 24,-

Vol. 186: Recent Trends in Graph Theory. Edited by M. Capobianco/ J. B. Frechen/M. Krolik. VI, 219 pages. 1971. DM 18.-

Vol. 187: H. S. Shapiro, Topics in Approximation Theory. VIII, 275 pages. 1971. DM 22,-

Vol. 188: Symposium on Semantics of Algorithmic Languages. Edited by E. Engeler. VI, 372 pages. 1971. DM 26,

Vol. 189: A. Weil, Dirichlet Series and Automorphic Forms. V. 164 pages. 1971. DM 16,-

Vol. 190: Martingales. A Report on a Meeting at Oberwolfach, May 17-23, 1970. Edited by H. Dinges. V, 75 pages. 1971. DM 16,-

Vol. 191: Séminaire de Probabilités V. Edited by P. A. Meyer. IV, 372 pages. 1971. DM 26,-

Vol. 192: Proceedings of Liverpool Singularities - Symposium I. Edited by C. T. C. Wall. V, 319 pages. 1971. DM 24,-

Vol. 193: Symposium on the Theory of Numerical Analysis. Edited by J. Ll. Morris. VI, 152 pages. 1971. DM 16,-

Vol. 194: M. Berger, P. Gauduchon et E. Mazet. Le Spectre d'une Variété Riemannienne. VII, 251 pages. 1971. DM 22,-

Vol. 195: Reports of the Midwest Category Seminar V. Edited by J.W. Gray and S. Mac Lane.III, 255 pages. 1971. DM 22,-

Vol. 196: H-spaces - Neuchâtel (Suisse)- Août 1970. Edited by F. Sigrist, V, 156 pages. 1971. DM 16,-

Vol. 197: Manifolds - Amsterdam 1970. Edited by N. H. Kuiper. V, 231 pages. 1971. DM 20,-

Vol. 198: M. Hervé, Analytic and Plurisubharmonic Functions in Finite and Infinite Dimensional Spaces. VI, 90 pages. 1971. DM 16.-

Vol. 199: Ch. J. Mozzochi, On the Pointwise Convergence of Fourier Series. VII, 87 pages. 1971. DM 16,-

Vol. 200: U. Neri, Singular Integrals. VII, 272 pages. 1971. DM 22,-

Vol. 201: J. H. van Lint, Coding Theory. VII, 136 pages. 1971. DM 16,-

Vol. 202: J. Benedetto, Harmonic Analysis on Totally Disconnected Sets. VIII, 261 pages. 1971. DM 22,-

Vol. 203: D. Knutson, Algebraic Spaces. VI, 261 pages. 1971. DM 22,-

Vol. 204: A. Zygmund, Intégrales Singulières. IV, 53 pages. 1971. DM 16,-

Vol. 205: Séminaire Pierre Lelong (Analyse) Année 1970. VI, 243 pages. 1971. DM 20,-

Vol. 206: Symposium on Differential Equations and Dynamical Systems. Edited by D. Chillingworth. XI, 173 pages. 1971. DM 16,-

Vol. 207: L. Bernstein, The Jacobi-Perron Algorithm - Its Theory and Application. IV, 161 pages. 1971. DM 16,-

Vol. 208: A. Grothendieck and J. P. Murre, The Tame Fundamental Group of a Formal Neighbourhood of a Divisor with Normal Crossings on a Scheme. VIII, 133 pages. 1971. DM 16,-

Vol. 209: Proceedings of Liverpool Singularities Symposium II. Edited by C. T. C. Wall. V, 280 pages. 1971. DM 22,-

Vol. 210: M. Eichler, Projective Varieties and Modular Forms. III, 118 pages. 1971. DM 16,-

Vol. 211: Théorie des Matroïdes. Edité par C. P. Bruter. III, 108 pages. 1971. DM 16,-

Vol. 212: B. Scarpellini, Proof Theory and Intuitionistic Systems. VII, 291 pages. 1971. DM 24,-

Vol. 213: H. Hogbe-Nlend, Théorie des Bornologies et Applications. V, 168 pages. 1971. DM 18,-

Vol. 214: M. Smorodinsky, Ergodic Theory, Entropy. V, 64 pages. 1971. DM 16,-

Vol. 215: P. Antonelli, D. Burghelea and P. J. Kahn, The Concordance-Homotopy Groups of Geometric Automorphism Groups. X, 140 pages. 1971. DM 16,-

Vol. 216: H. Maaß, Siegel's Modular Forms and Dirichlet Series. VII, 328 pages. 1971. DM 20,-

Vol. 217: T. J. Jech, Lectures in Set Theory with Particular Emphasis on the Method of Forcing. V, 137 pages. 1971. DM 16,-

Vol. 218: C. P. Schnorr, Zufälligkeit und Wahrscheinlichkeit. IV, 212 Seiten 1971. DM 20,-

Vol. 219: N. L. Alling and N. Greenleaf, Foundations of the Theory of Klein Surfaces. IX, 117 pages. 1971. DM 16,-

Vol. 220: W. A. Coppel, Disconjugacy. V, 148 pages. 1971. DM 16,-

Vol. 221: P. Gabriel und F. Ulmer, Lokal präsentierbare Kategorien. V, 200 Seiten. 1971. DM 18,-

Vol. 222: C. Meghea, Compactification des Espaces Harmoniques. III, 108 pages. 1971. DM 16,-

Vol. 223: U. Felgner, Models of ZF-Set Theory. VI, 173 pages. 1971. DM 16,-

Vol. 224: Revètements Etales et Groupe Fondamental. (SGA 1). Dirigé par A. Grothendieck XXII, 447 pages. 1971. DM 30,-

Vol. 225: Théorie des Intersections et Théorème de Riemann-Roch. (SGA 6). Dirigé par P. Berthelot, A. Grothendieck et L. Illusie. XII, 700 pages. 1971. DM 40,-

Vol. 226: Seminar on Potential Theory, II. Edited by H. Bauer. IV, 170 pages. 1971. DM 18,-

Vol. 227: H. L. Montgomery, Topics in Multiplicative Number Theory. IX, 178 pages. 1971. DM 18,-

Vol. 228: Conference on Applications of Numerical Analysis. Edited by J. Ll. Morris. X, 358 pages. 1971. DM 26,-

Vol. 229: J. Väisälä, Lectures on n-Dimensional Quasiconformal Mappings. XIV, 144 pages. 1971. DM 16,-

Vol. 230: L. Waelbroeck, Topological Vector Spaces and Algebras. VII, 158 pages. 1971. DM 16,-

Vol. 231: H. Reiter, L^1-Algebras and Segal Algebras. XI, 113 pages. 1971. DM 16,-

Vol. 232: T. H. Ganelius, Tauberian Remainder Theorems. VI, 75 pages. 1971. DM 16,-

Vol. 233: C. P. Tsokos and W. J. Padgett. Random Integral Equations with Applications to Stochastic Systems. VII, 174 pages. 1971. DM 18,-

Vol. 234: A. Andreotti and W. Stoll. Analytic and Algebraic Dependence of Meromorphic Functions. III, 390 pages. 1971. DM 26,-

Vol. 235: Global Differentiable Dynamics. Edited by O. Hájek, A. J. Lohwater, and R. McCann. X, 140 pages. 1971. DM 16,-

Vol. 236: M. Barr, P. A. Grillet, and D. H. van Osdol. Exact Categories and Categories of Sheaves. VII, 239 pages. 1971, DM 20,-

Vol. 237: B. Stenström. Rings and Modules of Quotients. VII, 136 pages. 1971. DM 16,-

Vol. 238: Der kanonische Modul eines Cohen-Macaulay-Rings. Herausgegeben von Jürgen Herzog und Ernst Kunz. VI, 103 Seiten. 1971. DM 16,-

Vol. 239: L. Illusie, Complexe Cotangent et Déformations I. XV, 355 pages. 1971. DM 26,-

Vol. 240: A. Kerber, Representations of Permutation Groups I. VII, 192 pages. 1971. DM 18,-

Vol. 241: S. Kaneyuki, Homogeneous Bounded Domains and Siegel Domains. V, 89 pages. 1971. DM 16,-

Vol. 242: R. R. Coifman et G. Weiss, Analyse Harmonique Non-Commutative sur Certains Espaces. V, 160 pages. 1971. DM 16,-

Vol. 243: Japan-United States Seminar on Ordinary Differential and Functional Equations. Edited by M. Urabe. VIII, 332 pages. 1971. DM 26,-

Vol. 244: Séminaire Bourbaki - vol. 1970/71. Exposés 382-399. IV, 356 pages. 1971. DM 26,-

Vol. 245: D. E. Cohen, Groups of Cohomological Dimension One. V, 99 pages. 1972. DM 16,-